Scientific Collaboration and Collective Knowledge

Scientific Collaboration and Collective Knowledge

New Essays

EDITED BY THOMAS BOYER-KASSEM
CONOR MAYO-WILSON
and
MICHAEL WEISBERG

OXFORD
UNIVERSITY PRESS

OXFORD
UNIVERSITY PRESS

Oxford University Press is a department of the University of Oxford. It furthers the University's objective of excellence in research, scholarship, and education by publishing worldwide. Oxford is a registered trade mark of Oxford University Press in the UK and in certain other countries.

Published in the United States of America by Oxford University Press
198 Madison Avenue, New York, NY 10016, United States of America.

Library of Congress Cataloging-in-Publication Data
Names: Boyer-Kassem, Thomas, editor.
Title: Scientific collaboration and collective knowledge / edited by Thomas Boyer-Kassem, Conor Mayo-Wilson, and Michael Weisberg.
Description: New York : Oxford University Press, 2018. | Includes bibliographical references and index.
Identifiers: LCCN 2016047948 (print) | LCCN 2017025590 (ebook) |
ISBN 9780190680541 (updf) | ISBN 9780190680558 (ebook) | ISBN 9780190680565 (online course) | ISBN 9780190680534 (hardback)
Subjects: LCSH: Science–International cooperation. |
Science–Research–International cooperation.
Classification: LCC Q172.5.I5 (ebook) | LCC Q172.5.I5 S35 2017 (print) | DDC 507.2–dc23
LC record available at https://lccn.loc.gov/2016047948

CONTENTS

LIST OF CONTRIBUTORS

Staffan Angere is an independent researcher who received his Philosophy PhD from Lund University, Sweden. He is the principal developer of the open source computer environment Laputa for doing simulations in social epistemology. He has published on philosophical logic, the philosophy of mathematics, epistemology, and the methodology of philosophy. Further research interests include logical empiricism, scientific methods for doing metaphysics, and general philosophy of science.

Denis Bonnay is an Assistant Professor at Université Paris Nanterre, where he is in charge of the master program in Philosophy. He has a PhD in Philosophy of Science from Université Panthéon Sorbonne. His research primarily pertains to formal philosophy, in areas ranging from philosophy of logic to social epistemology. He has proposed a new demarcation criteria for logical operations, and he works on extending this account of logicality outside deductive logic to systems of inductive reasoning and operations of descriptive statistics.

Justin Bruner is a Lecturer of Politics and International Relations at the Australian National University. He received his PhD in Philosophy from the University of California, Irvine in 2014, before which he studied economics and philosophy at the University of California, Berkeley. His research spans the fields of social and political philosophy, social and formal epistemology and the philosophy of biology. He is particularly interested in using mathematical models and experimental methods to grapple with philosophical questions and issues.

Bryce Huebner is an Associate Professor in the Philosophy Department at Georgetown University. He is the author of *Macrocognition: A Theory of Distributed Minds and Collective Intentionality* and the editor of *The Philosophy of Daniel Dennett*. His research focuses on group cognition, moral psychology, and issues in the philosophy of the social and cognitive sciences.

Rebecca Kukla is Professor of Philosophy and Senior Research Scholar in the Kennedy Institute of Ethics at Georgetown University, and a graduate student in geography at CUNY-Hunter College. Her books include *'Yo' and 'Lo': The Pragmatic Topography of the Space of Reasons* (with Mark Lance); *Mass Hysteria: Medicine, Culture, and*

Mothers' Bodies; and *City Living: How Urban Spaces and Urban Dwellers Make One Another* (forthcoming). She works on social epistemology, philosophy of applied science, anti-oppression theory, and philosophy of language.

Carlo Martini is a postdoctoral researcher in Philosophy of the Social Sciences at the Academy of Finland Centre of Excellence in the Philosophy of the Social Sciences (TINT, University of Helsinki, Finland). Carlo has worked and published numerous articles and a collected volume on the philosophy of expertise. His current work focuses on understanding how expert knowledge can be used as evidence in structured deliberation and how to build trust in scientific expertise.

Ryan Muldoon is Assistant Professor of Philosophy at the State University of New York at Buffalo. His primary interest is in how societies can benefit from being more diverse. He has published a number of articles in the social epistemology of science, social norms, and political philosophy. He recently published *Social Contract Theory for a Diverse World: Beyond Tolerance* (Routledge).

Cailin O'Connor (BA Harvard University, PhD University of California, Irvine) is an Assistant Professor in the Department of Logic and Philosophy of Science at UC Irvine. Her work is in philosophy of biology and the behavioral sciences, philosophy of science, evolutionary game theory, and formal social/feminist epistemology. She is working on a monograph tentatively titled *Dynamics of Inequity* and is currently administering the NSF grant Social Dynamics and Diversity in Epistemic Communities. Her Erdos-Bacon number is 8.

Erik J. Olsson is Professor and Chair in Theoretical Philosophy at Lund University, Sweden. His areas of research include epistemology, philosophical logic, pragmatism, and, more recently, philosophy of the Internet. Olsson has published extensively on epistemic coherence, the value of knowledge, social epistemology, and belief revision. Recent books include *Against Coherence: Truth, Probability, and Justification* (Oxford University Press, 2005), *Knowledge and Inquiry: Essays on the Pragmatism of Isaac Levi* (Cambridge University Press, 2006) and *Belief Revision Meets Philosophy of Science* (Springer, 2011). Olsson is co-developer of the computer environment Laputa for doing simulations in social epistemology.

Jan Sprenger is Professor of Philosophy at the School of Humanities at Tilburg University and Scientific Director of the Tilburg Center for Logic, Ethics and Philosophy of Science (TiLPS). He received his PhD from the University of Bonn in 2008 and has been working in Tilburg since then, mainly in philosophy of science, (formal) epistemology, and related areas. His papers are regularly published in journals such as *Mind*, *Philosophy of Science*, and *The British Journal for the Philosophy of Science*. Since 2015, he has conducted an ERC Starting Investigator project on the topic "Making Scientific Inferences More Objective."

Michael Strevens is Professor of Philosophy at New York University. His academic work is principally concerned with the nature of science, covering topics such as scientific explanation, understanding, complex systems, probability of various sorts, causation, and the social structure of science. A related set of projects uses contemporary work in cognitive psychology to explain aspects of both philosophical and scientific thinking.

Eric Winsberg is Professor of Philosophy at the University of South Florida. His principal research interests are in the philosophy of science, the philosophy of physics, the philosophy of climate science, and the role of models and simulations in science, generally. He is the author of *Science in the Age of Computer Simulation* and is currently working on the monograph *Philosophy and Climate Science*, which will appear in 2018.

K. Brad Wray joined the Centre for Science Studies at Aarhus University, in Denmark, in July 2017. He has published numerous articles on the social epistemology of science, collective intentionality, the anti-realism/realism debate, and Kuhn's philosophy of science. His book, *Kuhn's Evolutionary Social Epistemology*, was published by Cambridge University Press in 2011. He is one of the two co-editors of the Springer journal *Metascience*.

Kevin J. S. Zollman is an Associate Professor of Philosophy at Carnegie Mellon University and the author of *The Game Theorist's Guide to Parenting* (Scientific American/FSG, 2016). Zollman's research focuses on the use of mathematical models of social behavior in both humans and animals. Most recently, he has analyzed the effect of incentives on the social organization of science.

INTRODUCTION

In 2001, nearly three thousand scientists coauthored a single paper for the journal *Nature*. Although the sheer number of coauthors was unusual, the collaborative nature of the project from which the paper stemmed was not: over the last century, the size, complexity, and frequency of collaboration in the sciences have increased rapidly. For example, hundreds of scientists contribute to the intricate climate models from which environmental policy is derived. No single researcher has the empirical knowledge, mathematical training, and programming abilities to construct the models of clouds, precipitation, glacier movement, and so on, that are employed in such climate studies. Similar remarks apply to large-scale projects in particle physics and the biological sciences including, for instance, the human genome project, which produced the above-mentioned paper in *Nature*. Not less importantly, collaboration is not instantiated in very large groups only: small collaborations are ubiquitous in the sciences. Overall, single-author papers have become uncommon nowadays (Wuchty et al., 2007).

Despite its growing prevalence and importance for decades, only in the last few years has there been a growth in philosophical work analyzing collaborative research in the sciences.[1] What are the benefits and costs of such collaborations, and are current practices for encouraging collaborations optimal? How should credit for discovery and responsibility for error be attributed to members of a lab or coauthored paper? How ought collaborating scientists summarize their findings if they disagree about the interpretation of their results? The chapters in this volume attack these difficult questions, among others.

Although some contributors employ empirical research,[2] the ensuing articles contain no new data from journal databases, no new surveys of scientists, and no novel analyses of current trends in memberships to scientific societies. Instead, contributions attack *conceptual* questions about the ways in which scientific collaborations might take place and *normative* questions about how scientific institutions ought to be organized in light of such possibilities. For this reason, this volume

[1] Some recent works include Andersen (2014a,b), Andersen and Wagenknecht (2013), Boyer-Kassem and Imbert (2015), Fallis (2006), Frost-Arnold (2013), Galison et al. (2003), Rolin (2015), Thagard (2005), Winsberg et al. (2014), Wray (2002, 2006), Wylie (2014), and Wylie (2015).

[2] See Bruner and O'Connor's contribution, for instance.

is a natural extension of research in the philosophy of science and the growing field called "social epistemology." Although primarily philosophical, the questions addressed in this volume are important not only for philosophers but also for scientific policymakers and for social scientists who are interested in what data *ought* to be collected to advance our understanding of scientific practice and how it ought to be improved.

How do contributors address these conceptual and normative questions? Several authors represent scientific communities via mathematical models, and they study these models either analytically (Bonnay and Strevens) or through computer simulations (Angere and Olsson, Bruner and O'Connor, and Zollman). Other authors (Huebner, Kukla and Winsberg, and Wray) provide detailed analysis of case studies of past and current scientific collaborations, thereby highlighting the complexities of contemporary scientific practice. Many contributors (especially Bruner and O'Connor, Muldoon, and Martini and Sprenger) employ a somewhat hybrid methodology of analyzing large-scale, statistical trends in scientific practice and then proposing abstract models to explain those trends. All contributors, however, share the method of employing normative theories from epistemology and philosophy of science to draw morals about existing collaborative research, possible patterns of collaboration, and how scientific institutions can be improved.

The virtues and vices of case-study analysis, formal modeling, and philosophical analysis are well-known.[3] Here, we briefly discuss the virtues of these methods as they are employed by authors in this volume. One virtue of the formal models in this volume is their generality: they apply to many types of collaborations, even ones among non-scientists. For instance, Strevens presents a "waiting-time model of discovery" that assumes very little about how often scientists make discoveries. In principle, the model could be applied to any group engaged in inquiry, empirical or not. As a second example, Bruner and O'Connor's model explains why minorities might receive less credit for their contributions to coauthored scientific papers, but there is nothing in principle that prohibits their model from being applied to cases of credit-sharing in collaborations in the humanities.

Because these formal models are quite general, they do not represent many important features of scientific communities. The case studies investigated in the Part III of this volume in contrast, are extremely rich and highlight the difficulties of assigning credit and blame in "radically collaborative" projects in science. Huebner, Kukla, and Winsberg consider the case of modular computer simulations in contemporary climate science, as well as multisite clinical trials in biomedical research, in which no single scientist can vouch for the reliability of all the group's findings. Wray studies policies of medical journals and of the National Science Foundation. A central virtue of these case studies is that, by illustrating the complexity of large-scale scientific collaborations, they indicate the inadequacy of existing concepts for describing scientific research and the limitations of existing norms for assigning credit for discovery and blame for

[3] On case study analysis, see Burian (2001), Chang (2011), Currie (2015), Kinzel (2015), Morgan (2012), and Pitt (2001). On the use of formal models, see Salmon (1978), Salmon and Salmon's (1979) criticisms of formal models in the social sciences, for example. For a defense of mathematical philosophy, see for instance Leitgeb (2013).

misconduct. These contributions, therefore, force us to extend our current vocabulary and norms to understand and improve contemporary scientific practice.

Regardless of methodology, contributors to this volume study scientific groups of various scales and natures: a "group" might refer to the whole scientific community, a research program, a university department, a peer-review board, or a specific collaboration of a few scientists. Huebner, Kukla, and Winsberg even discuss cases in which non-academics are involved in scientific collaborations.[4] Thus, the contributions to the volume take a first step toward understanding the rich philosophical issues surrounding collaboration that arise at all levels of scientific practice, from data collection to abstract theorizing.

This volume is part of the larger "social turn" in philosophy of science and epistemology.[5] Throughout much of the twentieth century, many philosophers of science studied properties of scientific theories and neglected the social nature of science. Central research topics included explanation, confirmation, probability, and causation, for instance. Similarly, although epistemologists developed thought experiments with human protagonists, these philosophers were interested in evidence, justification, and knowledge in the abstract, potentially divorced from any human subject. In both epistemology and philosophy of science, epistemic norms, either for the everyday individual or for the working scientist, were often understood as descriptions of the mental life of an isolated, stoic, ideally rational agent.

Starting with Kuhn, the focus on the individual, ideally rational scientists started to erode. In the ensuing decades, philosophers of science and epistemologists began to investigate the serious philosophical issues that arise from the social nature of science. Collective knowledge (Nelson, 1990), trust within large scientific communities (Hardwig, 1985, 1991), and testimony (Lackey and Sosa, 2006) became central issues of philosophical concern. As the contributions to this volume show, investigating the practices of collaboration in sciences raises a new, rich set of philosophical questions about communication, collective responsibility, authorship, and more.

The chapters of the volume tackle various problems in relation to scientific collaboration and group knowledge. We summarize the key points for each part of the volume.

Part I: Sharing Knowledge

What should scientists communicate to which others? This is the central question of Part I of this volume. Strevens invites us to consider a radical strategy that stems from the "communist" norm of the sociologist of science Merton: share everything with

[4] Collaboration among scientists and non-academics is becoming more common in archaeology, where participation of indigenous communities has been tremendously fruitful. See Wylie (2014, 2015).

[5] Longino (2006) recognizes, in fact, that it is better to call this movement a "social *re*turn" to emphasize that the early logical positivists had likewise emphasized the social nature of science. For example, Hahn et al. (1929) write: "The scientific world conception is characterized not so much by theses of its own, but rather by its basic attitude, its points of view and direction of research ... The endeavor is to link and harmonize the achievements of individual investigators in their various fields of science. From this aim follows the emphasis on collective efforts."

everyone. Angere and Olsson consider the alternative proposal that scientists ought to limit their communications to information that is above some threshold in quality.

Strevens's surprising conclusion is that Hobbesian scientists, who only care for their individual benefit, should be willing to sign a contract for total sharing of "all pre-publication information unconditionally," and that this contract is the best possible one on information sharing. Thus, he provides a "Hobbesian vindication" of Merton's communist norm, which partially explains why the norm does and should prevail. Angere and Olsson, in contrast, vindicate the intuition that it is in general better for inquirers to refrain from sharing low-quality information, which "pulls the entire research network away from truth." On the other hand, their finding that there are cases in which not communicating at all prior to voting is the best option may strike the reader as surprising.[6]

These chapters raise a number of questions for future research. Let us mention a few ones, in addition to the ones that the authors already indicate. Strevens assumes that sharing information is free. Thus, it would be interesting to investigate whether Strevens's results still hold if information retrieval were costly, as the models of Angere, Olsson, and Zollman suppose. Another question for Strevens's important result is how the Hobbesian contract can be implemented in practice. How should we respond to free-riders, who benefit from the sharing of others, but do not share themselves? How could or should the contract be enforced?

Angere and Olsson assume that each scientist's sources (i.e., the other individuals in the research network) are independent. This is obviously an idealization. Scientists within a common research community share methods and knowledge—sometimes even raw data or material. Consequently, sources of error and evidence are likely to be correlated in some sense. So the next step coul be to study how Angere and Olsson's model extends to cases in which informational sources are dependent.

Part II: Forming Collaborations

Why do scientists collaborate and how can we encourage more fruitful collaborations? This is the central question of the chapters by Muldoon and Zollman.

Muldoon begins by discussing a limitation of some existing mathematical models of the "division of cognitive labor" in science: paradoxically, some models fail to recognize that scientists have different skill sets and that researchers often choose projects based upon what skills they currently have. Muldoon argues that by combining this simple fact with the observation that acquiring skills is costly, we can better explain the rapid rise of collaboration in the sciences, the social structure of evolving collaborative networks, and new trends in science, such as disciplinary "colonization." As an example of "colonization," Muldoon explores how physicists are in increasing demand in biology, where the mathematical training of physicists is in high demand.

[6] Arguments for such a conclusion are not new, however. In Book III (Chapter 2) of *Of Social Contract*, Rousseau argues, "If, when the people deliberates and is adequately informed, the citizens were to have no private communication among themselves, the general will would always result from the large number of small differences and the deliberation would always be good" (Rousseau 2012). Similar conclusions have been defended on the basis of Condorcet's jury theorem.

Zollman takes a first step in addressing Muldoon's challenge to develop a formal model in which scientists have different skill sets and different information. He assumes that scientists possess different "conceptual schemes," and, in order to solve a given problem, scientists may need to employ the schemes of others. For simplicity, Zollman models "one-way" collaboration, in which one scientist borrows another's conceptual scheme at a cost. Given this setting, Zollman characterizes the optimal structure for a scientific community and argues that, surprisingly, it can be counter-productive to encourage scientists to seek out new collaborators, as doing so may lead to failure of certain conceptual schemes to be transferred widely.

Muldoon's and Zollman's arguments raise interesting issues for philosophers and scientific policymakers alike. For example, in Zollman's model of collaboration, scientists share their own way of doing research voluntarily. But as Strevens discusses, there are sometimes strong incentives against sharing information in the sciences. So how would Zollman's results be changed if a system of incentives were created for the scientists to share their conceptual schemes? And if Muldoon is right that disciplinary "colonization" is unavoidable, how could scientific institutions best foster fruitful collaborations for researchers who move into scientific fields with very different conceptual schemes?

Part III: Authorship and Refereeing in Collaborative Research

The third part of the volume investigates authorship, refereeing, credit, and responsibility in collaborative scientific research. When several scientists sign a paper, is each of them an author in the same sense they would be for a single-authored paper? Do each of the authors, or at least one, have to agree with each claim made in the paper? Or can some claims be endorsed only collectively, and by no one individually? If credit is shared for coauthored papers, what would a fair sharing between collaborators be? Or, how would other credit-sharing rules give incentives for other patterns of collaboration? These are the types of questions addressed by Huebner, Kukla, and Winsberg, Wray, and Bruner and O'Connor.

Huebner, Kukla, and Winsberg distinguish between collaborative research and multiple authorship, and put forth the case of radically collaborative research, "that is distributed widely over disciplinary expertise, time, and space," like in some climate-modeling projects that are modular in their development. They argue that this distributed collaboration threatens our common understanding of authorship because many of the methodological choices made by scientists cannot be checked by their collaborators, and yet those play an indispensable role in the final result. The work by Huebner, Kukla, and Winsberg suggests considering the roles of values in shaping these choices, making a link with classical discussions on this topic.[7]

Like Huebner, Kukla, and Winsberg, Wray considers the question of the adequacy of the traditional notion of authorship in collaborative research, for instance in cases where hundreds of scientists coauthor a paper. After analyzing the authorship criteria that some institutions and journals have recently adopted, he criticizes the requirement that authors identify their contributions. Finally, Wray investigates

[7] See Douglas (2009) and Rudner (1953).

whether current refereeing practice is well-suited for (large) collaborative works. If the skills of hundreds of scientists were necessary to produce some piece of research, and if no individual scientist can justify the group's conclusions, what kind of assessment can an individual referee provide? Wray suggests that one might consider referee teams, composed of individuals with various skills that enable them to *collectively* evaluate a paper or a grant proposal.

Bruner and O'Connor investigate authorship and credit in increasingly diverse scientific research teams. Substantial research has defended the value of *methodological* and *ideological* diversity.[8] That is, there is ample empirical and theoretical evidence that when scientists address the same questions using different methods, background theories, and values, their results are more accurate and serve the interests of a greater number of scientists in their communities. Diverse ideas and methods, of course, often originate from diverse people; such is the feminist insight that the inclusion of minorities in research improves the products of science.[9] Nonetheless, few philosophers have studied how diversity might affect the attribution of credit in scientific communities with power inequalities. Bruner and O'Connor tackle this difficult issue. Using several evolutionary game-theoretic models, they show why minorities might receive systematically less credit for their contributions if they must negotiate with collaborators for first-author or principal-investigator status. They find similar results for collaborations within existing power hierarchies, like those between a senior researcher and her graduate students. Their results explain existing empirical findings about inequality in first-authorship status, and, perhaps as equally important, their results point the way to rectify the problem.

All three chapters in Part III raise important concerns that the traditional notion of authorship faces in the age of collaboration. They consider possible remedies, yet problems remain. Thus, journals and editorial boards would benefit from further investigation of the best refereeing practices for collaborative works. More generally, given the role credit and citation has acquired in contemporary science (see e.g., Strevens, 2003), a wider critical appraisal of the reward system of science seems unavoidable as collaboration increases. Regarding Bruner's and O'Connor's contribution, a question that could be posed is to what extent the proposed models are realistic and represent actual features of academic research. The authors rightly note that the models exhibit *possible* mechanisms that could be looked for by sociologists. But even if these mechanisms were not found, the models could actually be of some explanatory value, as it is well-known from philosophy of science debates that idealized or unrealistic models are not valueless. More generally, and this holds for the models proposed in other chapters too, an analysis of the explanatory power of the simplified models that are used would be a welcome future addition.

[8] See Hong and Page (2001), Kitcher (1990), Longino (1990, 2001), Mayo-Wilson et al. (2011), Weisberg and Muldoon (2009), Wylie (2012), and Zollman (2010).

[9] See Intemann (2009, 2010), Longino (1990, 2001), and Wylie (2003, 2012).

Part IV: From Individual to Collective Opinion

The last two chapters in the volume address the question: How are and ought a scientific community's judgments be related to those of individual scientists? Martini and Sprenger focus on the particular problem of characterizing the best methods for amalgamating an expert community's estimates of a single quantity. For example, imagine a panel of nuclear scientists and medical researchers is asked to estimate the maximum amount of radiation to which one can be safely exposed. Or suppose climate scientists are asked how many additional hurricanes there will be in fifty years due to climate change. In these cases, individual scientists' estimates will likely differ. How ought we combine their estimates in order to decide what to believe about the quantity or to make a policy?

Martini and Sprenger consider two types of strategies: (1) treat all scientists as equally reliable and take a straight average of their guesses, and (2) try to weight expert estimates by perceived reliability. Employing mathematical results of Klein and Sprenger (2015), Martini and Sprenger discuss the conditions under which treating experts as differently reliable leads to better estimates, even if we are not certain how exactly reliable different experts are. Nonetheless, the mathematical results they discuss require that we can *compare* expert reliability, and so Martini and Sprenger discuss the empirical prospects for accurate comparisons. They find modest support for the claim that, in limited circumstances, comparisons of expert reliability are possible, but often, equal weighting is just as effective in practice.

Martini and Sprenger largely ignore the question of identifying the *reasons* expert estimates differ. Suppose that instead of aggregating expert judgments, a scientific institution or grant-giving body (e.g., the National Science Foundation in the United States) wishes to identify different scientific camps to discern why there is disagreement about some hypothesis. How might the institution identify *clusters* of scientists who share similar theoretical commitments? Bonnay's paper addresses this important question.

Bonnay first discusses axioms that any procedure for dividing scientists into clusters ought to satisfy. For example, cloning a scientist might increase the size of a cluster in an existing scientific community (namely, the cluster with her clone), but it should not result in the creation of new clusters; the clone does not add any ideological diversity to the existing community. Bonnay then proves that all clustering methods satisfying a few axioms have a similar functional form. Bonnay's work, therefore, not only provides a principled method for identifying like-minded groups of scientists, but it also provides insight into difficult metaphysical questions about when a group has a shared view or purpose.

Bonnay's formal framework differs from that of Martini and Sprenger in an important way. Whereas Martini and Sprenger represent scientists' beliefs using numbers (which represent a scientist's estimate of some unknown quantity), Bonnay represents beliefs using sets of propositions. This difference raises interesting prospects for future research for their respective projects. For example, can Bonnay's principles concerning propositional beliefs be integrated with existing axioms for dividing groups into clusters based on their quantitative beliefs (Ben-David and Ackerman, 2009)?

Thomas Boyer-Kassem, Conor Mayo-Wilson, and Michael Weisberg

References

Andersen, Hanne. Co-author responsibility. *EMBO reports*, 15(9):914–918, 2014a.

Andersen, Hanne. Epistemic dependence in contemporary science: Practices and malpractices. In Lena Soler, Sjoert Zwart, Michael Lynch, and Vincent Israel-Jost, editors, *Science after the practice turn in the philosophy, history and social studies of science*, volume 14 of *Routledge Studies in the Philosophy of Science*. Routledge, 2014b.

Andersen Hanne and Susann Wagenknecht. Epistemic dependence in interdisciplinary groups. *Synthese*, 190(11):1881–1898, 2013.

Ben-David Shai and Margareta Ackerman. Measures of clustering quality: A working set of axioms for clustering. In *Advances in neural information processing systems*, 121–128, 2009. URL http://papers.nips.cc/paper/3491-measures.

Boyer-Kassem Thomas and Cyrille Imbert. Scientific collaboration: Do two heads need to be more than twice better than one? *Philosophy of science*, 82(4):667–688, 2015.

Burian Richard M. The dilemma of case studies resolved: The virtues of using case studies in the history and philosophy of science. *Perspectives on Science*, 9(4):383–404, 2001.

Chang Hasok. Beyond case-studies: History as philosophy. In Seymour Mauskopf and Tad Schmaltz, editors, *Integrating history and philosophy of science*, number 263 of *Boston Studies in the Philosophy of Science*, 109–124. Springer, 2011. doi 10.1007/978-94-007-1745-9_8.

Currie Adrian. Philosophy of science and the curse of the case study. In Christopher Daly, editor, *The Palgrave handbook of philosophical methods*, 553–572. Springer, 2015.

Douglas Heather. *Science, policy, and the value-free ideal*. University of Pittsburgh Press, 2009.

Fallis Don. The epistemic costs and benefits of collaboration. *The Southern Journal of Philosophy*, 44(S1):197–208, 2006.

Friedrich Stadler, and Thomas Uebel (eds.), 2012, Wissenschaftliche Weltauffassung. Der Wiener Kreis. Reprint of original Edition with Translations, Vienna: Springer.

Frost-Arnold Karen. Moral trust & scientific collaboration. *Studies in History and Philosophy of Science Part A*, 44(3):301–310, 2013.

Galison, Peter. The collective author. In Peter Galison and Mario Biagioli, editors, *Scientific authorship: Credit and intellectual property in science*, 325–353. Routledge, 2003.

Hardwig Jonathan. Epistemic dependence. *The Journal of Philosophy*, 82(7):335–349, 1985.

Hardwig Jonathan. The role of trust in knowledge. *The Journal of Philosophy*, 88(12):693–708, 1991.

Hong Luo and Scott E. Page. Problem solving by heterogeneous agents. *Journal of Economic Theory*, 97(1):123–163, 2001.

Intemann Kristen. Why diversity matters: Understanding and applying the diversity component of the National Science Foundation's broader impacts criterion. *Social Epistemology*, 23 (3-4):249–266, 2009.

Intemann Kristen. 25 years of feminist empiricism and standpoint theory: Where are we now? *Hypatia*, 25(4):778–796, 2010.

Kinzel Katherina. Narrative and evidence. How can case studies from the history of science support claims in the philosophy of science? *Studies in History and Philosophy of Science Part A*, 49:48–57, February 2015. ISSN 0039-3681. doi: 10.1016/j.shpsa.2014.12.001.

Kitcher Philip. The division of cognitive labor. *The Journal of Philosophy*, 87(1):5–22, 1990.

Klein Dominik and Jan Sprenger. Modelling individual expertise in group judgements. *Economics and Philosophy*, 31(01):3–25, 2015.

Lackey Jennifer and Ernest Sosa. *The epistemology of testimony*. Clarendon Press, 2006.

Leitgeb Hannes. Scientific philosophy, mathematical philosophy, and all that. *Metaphilosophy*, 44(3):267–275, 2013.

Longino Helen E. *Science as social knowledge: Values and objectivity in scientific inquiry*. Princeton University Press, 1990.

Longino Helen E. *The fate of knowledge*. Princeton University Press, 2001.

Longino Helen E. Philosophy of science after the social turn. In Maria Carla Galavotti, editor, *Cambridge and Vienna*, number 12 in Vienna Circle Institute Yearbook (2004), 167–177. Springer, 2006.

Mayo-Wilson Conor, Kevin J. Zollman, and David Danks. The independence thesis: When individual and social epistemology diverge. *Philosophy of Science*, 78(4):653–677, 2011.

Morgan Mary S. Case studies: One observation or many? Justification or discovery? *Philosophy of Science*, 79(5):667–677, 2012.

Nelson Lynn. *Who knows: From Quine to a feminist empiricism*. Temple University Press, 1990.

Pitt Joseph C. The dilemma of case studies: Toward a Heraclitian philosophy of science. *Perspectives on Science*, 9(4):373–382, 2001.

Rolin Kristina. Values in science: The case of scientific collaboration. *Philosophy of Science*, 82(2): 157–177, 2015.

Rousseau Jean-Jacques. *The major political writings of Jean-Jacques Rousseau: The two "discourses" and the "social contract,"* University of Chicago Press, 2012.

Rudolf Carnap, Hans Hahn, and Otto Neurath, 1929, *Wissenschaftliche Weltauffassung – Der Wiener Kreis*, Wien: Wolf, trans. in Stadler and Uebel (2012), 75–116.

Rudner Richard. The scientist qua scientist makes value judgments. *Philosophy of science*, 20(1): 1–6, 1953.

Salmon Merrilee H. What can systems theory do for archaeology? *American Antiquity*, 43(2): 174–183, 1978. doi: 10.2307/279242.

Salmon Merrilee H and Wesley C. Salmon. Alternative models of scientific explanation. *American Anthropologist*, 81(1):61–74, 1979.

Strevens Michael. The role of the priority rule in science. *The Journal of Philosophy*, 100(2):55–79, 2003.

Thagard Paul. How to collaborate. *The Southern Journal of Philosophy*, 44(Supplement):177–196, 2005.

Weisberg Michael and Ryan Muldoon. Epistemic landscapes and the division of cognitive labor. *Philosophy of Science*, 76(2):225–252, April 2009.

Winsberg Eric, Bryce Huebner, and Rebecca Kukla. Accountability and values in radically collaborative research. *Studies in History and Philosophy of Science Part A*, 46:16–23, 2014.

Wray, Brad K. The epistemic significance of collaborative research. *Philosophy of Science*, 69(1): 150–168, 2002.

Wray, Brad K. Scientific authorship in the age of collaborative research. *Studies in History and Philosophy of Science Part A*, 37(3):505–514, 2006.

Wuchty Stefan, Benjamin F. Jones, and Brian Uzzi. The increasing dominance of teams in production of knowledge. *Science*, 316(5827):1036–1039, 2007.

Wylie Alison. Why standpoint matters. *Science and other cultures: Issues in philosophies of science and technology*, 26–48, 2003.

Wylie Alison. Standpoint matters. In *Proceedings and Addresses of the American Philosophical Association*, 86.2 (2012): 47–76.

Wylie Alison. Community-based collaborative archaeology. In Nancy Cartwright and Elenora Montuschi, editors, *Philosophy of social science: A new introduction*, 68–82. Oxford University Press, 2014.

Wylie Alison. A plurality of pluralisms: Collaborative practice in archaeology. In Flavia Padovani, Alan Richardson, and Jonathan Y. Tsou, editors, *Objectivity in science*, number 310 of *Boston Studies in the Philosophy and History of Science*, 189–210. Springer, 2015.

Zollman Kevin J. The epistemic benefit of transient diversity. *Erkenntnis*, 72(1):17–35, 2010.

PART ONE

SHARING KNOWLEDGE

1

Scientific Sharing, Communism, and the Social Contract

MICHAEL STREVENS

1.1 Sharing, For and Against

Society has a clear and strong interest in the wide distribution of scientific knowledge. Such dissemination has direct and indirect social benefits. The direct benefits come through the public consumption of knowledge, in the form of miracle cures, magical gadgets, and eternal truths to contemplate. The indirect benefits come through other scientists' consumption of knowledge, since once digested such knowledge will tend to increase the rate, breadth, and depth of scientific discovery and so the magnitude of the direct benefits. (Qualifications should no doubt be made about weaponized anthrax, cobalt bombs, and so on.)

Scientists' own interest in sharing the knowledge they have themselves created is more circumscribed. Their desire to make the world a better place and, even if that desire is lacking, to receive credit for their discoveries militates in favor of sharing. But the threat of one scientist's finagling credit for another scientist's discovery militates, at least temporarily, in favor of secrecy.

The reward system in science—the "priority rule," according to which only the first scientist or team of scientists to make a discovery receives credit for the discovery (Merton 1957)—increases the strength of the arguments both pro and con. On the one hand, the priority system gives you reason to publish your results as early as possible, in order to maximize your chances of being recognized as the first discoverer (Dasgupta and David 1994). On the other hand, the same considerations give you a powerful incentive not to share your results before you have extracted every last publication from them, to minimize the chances that someone else takes advantage of your research to gain credit that might otherwise have gone to you.

These pro and con arguments apply to largely non-overlapping parts of the research and publication cycle. Suppose, somewhat artificially, that the two parts are entirely distinct, that is, that every research program can be divided into phases before and after discovery. The priority rule motivates a scientist to keep all data, all technology of experimentation, all incipient hypothesizing secret before discovery, and then to publish, that is to share widely, anything and everything of social value as soon as

Boyer-Kassem, Mayo-Wilson, & Weisberg. *Scientific Collaboration and Collective Knowledge*.

possible after discovery (should a discovery actually be made). The interests of society and the scientist are therefore in complete alignment after discovery, but before discovery, they appear to be diametrically opposed.[1]

The resulting moral and political conflicts are dramatically portrayed in case studies such as James Watson's (1968) account of the discovery of the structure of DNA, in which Watson wheedles a crucial X-ray diffraction image of the molecule from the Wilkins lab without the crystallographer Rosalind Franklin's knowledge, or Wade's (1981) account of Roger Guillemin and Andrew Schally's race to sequence the brain-signaling hormone TRH (Schally's policy on sharing: "Don't talk to the enemy!"), or Collins's (1974) account of the development of the TEA laser, in which rivalry among competing labs discourages communication that might have sped everyone to their joint goal—with one scientist memorably confessing to the following policy on sharing research: "Let's say I've always told the truth, nothing but the truth, but not the whole truth" (p. 180).

This is all quite entertaining, but it prompts a serious question. Are there circumstances, or can circumstances be created, under which information will be shared in a way that precipitates, rather than merely promulgating, discoveries? Let me begin by surveying some of science's current pro-sharing arrangements.

Certain grant agencies require a degree of data sharing as a condition for accepting their funds. In the United States, both the National Institutes of Health (the NIH) and the National Science Foundation (the NSF), two of the largest government grantors, have explored this approach. Typically, the data to be shared has been used as the basis for publications (there is no obligation to share before publication), but further significant publications might be extracted from the same body of information; this is the sort of case where the "before" and "after" phases distinguished above are intermingled. Some publicly minded publishers, such as PLOS, have similar policies.[2]

In special cases, notably in the life sciences, rules for the sharing of certain classes of data have been agreed upon by funding agencies, industry representatives, and the heads of research projects working together. An early and influential example was the negotiation in 1996 of the "Bermuda Principles" mandating the immediate publication of DNA sequences obtained by the Human Genome Project (Contreras 2011).

The power of such requirements is limited, however, so long as the organizations in question are fighting against individual researchers' own aims and desires: the NIH planned a demanding data-sharing protocol in the early 2000s, but this was watered down in 2003 after objections from many scientists.[3] Further, the publication trigger may come far later than is socially optimal, and in any case, data is only one

[1] Many writers have discussed the conflict between sharing and scientists' narrow self-interest; for a sample of approaches see Dasgupta and David (1987), Arzberger et al. (2004), Jasonoff (2006), and Resnik (2006).

[2] Though to sometimes limited effect (Savage and Vickers 2009).

[3] The "Final NIH Statement on Sharing Research Data" reads: "Several groups and individuals objected to sharing of research data prior to publication ... NIH recognizes that the investigators who collect the data have a legitimate interest in benefiting from their investment of time and effort. We have therefore revised our definition of 'the timely release and sharing' to be no later than the acceptance for publication of the main findings from the final data set. NIH continues to expect that

aspect of the epistemic and material goods generated by a research program's progress that may be fruitfully shared before discovery—think of software and computational methods, genetically engineered lab animals, the sort of instrumental or surgical know-how that has to be learned in person, the loan of equipment or technicians, and so on.

A more radical strategy is to abandon priority as a principle for allocating credit throughout science, thus removing much of the disincentive to share. A step in this direction has been taken by Nancy Wexler's Hereditary Disease Foundation:

> In larger projects which require many scientific lines of inquiry at once, HDF funded scientists work together toward the same end goal. Credit for breakthroughs is shared by all collaborators, regardless of which vein of research and scientist(s) "arrives first." The HDF community understands that complex problems are solved more rapidly when cooperation between those with complementary skills takes place. (Hereditary Disease Foundation 2014)

The priority rule, however, performs a valuable function in determining an efficient allocation of cognitive labor among scientific projects (Strevens 2003). It solves, in effect, an enormously complex and pressing coordination problem, and so—in spite of its often rough justice—it should be if at all possible retained.

Rather than forcing scientists to share their knowledge as a condition of their funding, then, it would be better to find ways to make them want to share, building if possible on preexisting inclinations. Are there, already ensconced in the scientific mind, any impulses toward the free dissemination of information?

A partial answer to this question is provided by Boyer's (2014) and Heesen's (2016) important arguments that it is rational for scientists solely concerned with the pursuit of personal credit to break up their work into "least publishable units" and to place these units in the journals as soon as possible, despite the collateral advantage it gives to their competitors. What looks altruistic—sharing early and widely—is thereby motivated by narrow self-interest, seldom in short supply. The benefits are circumscribed, however: as with data-sharing requirements, nothing will be shared until something relevant is ready for publication, and worse, it is only what characteristically goes into the journals that gets broadcast, so the other shareables mentioned above will remain hidden or otherwise unavailable.

Agreements to trade information for mutual benefit are another means of sharing. Like Boyer's and Heesen's atomized publication strategy, these contractual exchanges are pursued for private advantage; unlike that strategy, they are not confined to publishable information and goods.

All of this arm-twisting and self-regard is, however, nothing but an ugly and awkward prelude to a far sweeter song: an ideal of sharing already firmly established in science's guiding system of values.

the initial investigators may benefit from first and continuing use but not from prolonged exclusive use" (National Institutes of Health 2003).

1.2 The Communist Way

What Robert Merton provocatively but accurately called science's communist norm enjoins common ownership of scientific information: "secrecy is the antithesis of this norm; full and open communication its enactment" (Merton 1942, 274). To the extent that allegiance to the communist norm opposes the dictates of narrow self-interest, it drives sharing in science. When self-interest anyway points toward sharing, communism may help smooth the way with good feelings and the satisfying sense of duty done. (Merton focuses on information rather than on more tangible research-related goods such as bacterial strains or genetically modified mice. The communist norm appears to apply to some extent to these, but the boundaries are unclear. I follow Merton's lead in focusing on information, but you might also understand the term "information" in the remainder of this chapter as a catch-all for any kind of shareable.)

Merton offered little evidence for the existence of informational communism; he presumed, I suppose, that every scientist would recognize it as a part of their professional code. Later more empirically scrupulous sociological studies have vindicated his confidence. A survey conducted by Macfarlane and Cheng (2008) found 95% of respondents (of whom there were several hundred, mostly UK university professors) in favor of sharing research in progress. A similar result was found in a survey of nearly 2000 US geneticists and other life scientists, with 91% agreeing that they should "freely share information, data, or materials with all academic scientists" (Louis et al. 2002).[4] Even scientists working in private industry endorse unrestricted sharing, though they are according to the study somewhat less likely to realize the ideal in practice.

More evidence for the communist norm comes from the rhetoric surrounding the formulation of sharing agreements such as the Human Genome Project's Bermuda Principles: Contreras (2011) documents the importance of "strong open science norms" both in motivating the original agreement and in its subsequent implementation by other institutions (pp. 82, 88).

There is little reason, then, to question the existence of a communist norm. But the norm does not make scientists' decisions to share straightforward. The negative consequences of sharing are palpable: 35% of the surveyed life scientists reported that, as a result of sharing information, they had been "scooped" by other researchers, losing the opportunity to gain credit for work based on their own results. Perhaps unsurprisingly, then, scientists frequently fail to live up to the communist ideal. About 30% said that they had "within the past three years withheld research results from other academic scientists prior to publication on at least one occasion"—not unlike Collins's laser scientist quoted in the previous section. Subscribing to and complying with a norm are, as we all well know, not the same thing.[5]

[4] These studies make clear something about which Merton was vague: the moral imperative to share applies to unpublished as well as published research.

[5] Tenopir et al. (2011) provide some further statistics on the spotty sharing of data.

It may seem peculiar that the norm of communism exists at all, given scientists' powerful motivation to ignore its prescriptions. And you might wonder how long it can endure in the mercilessly competitive world of science today.[6]

The principal aim of this chapter is to provide what I will call a Hobbesian vindication of the communist norm, a vindication that might variously be used to justify, to shore up, or to explain the origin of the norm.

A Hobbesian vindication of a norm has two parts. The first part is a transformation of the norm into a social contract that behaviorally mirrors the norm. Rather than a moral obligation to share, then, we consider a contract in which the signer promises to behave just as the norm would oblige them: to share all pre-publication information unconditionally. (It is assumed in the Hobbesian spirit that the contract creates an irresistible, but in this chapter unspecified, means of enforcement.)

The second part is a rationale for signing the contract. Typically it is assumed that either everyone signs the contract—the "cooperation" scenario—or that no one does—the "state of nature." The cooperation scenario is shown to be superior, for everyone involved, to the state of nature. In the case of sharing, then, the aim is to demonstrate that it is better for all scientists if they share universally, as required by the communist norm, than if they do not share at all. An even stronger vindication would demonstrate that the communist contract is more desirable than any other sharing contract; that too I will attempt to show.

What is the point of a Hobbesian vindication of a norm? It concerns a fictional social contract rather than the actual moral precept in question; it tells (by implication) an origin story, a collective agreement to exit the state of nature, that is unlikely to have any precise historical parallel; and it has nothing practical to say about the means by which such an agreement might be negotiated or enforced.

Nevertheless, the vindication can be used, given the right sort of philosophical or empirical scaffolding, to do a number of things: to justify the norm, to give an evolutionary account of the origin of the norm, to give an account of the social or psychological stability of the norm, and as a tool for social engineering, by repairing or enhancing the stability of or (in some cases) dismantling the norm. With respect to the communist norm, I will not attempt any of this. But I take it that, by providing the Hobbesian vindication, I will have cast some light on the best way to go about achieving any or all of these aims.

Let me turn, then, to the question of contractual sharing. Consider scientists working in a near state of nature: no institutionalized agreements to share, no

[6] Humans are altruistic: they recognize a ceteris paribus obligation to take steps, when possible, to improve the world at large. Might the communist norm be explained, then, as an application of the general altruistic urge to the special case of unpublished scientific information? Might it be, not a norm particular to science, but a manifestation of the broadest moral norm of all?

Altruism certainly plays a role in motivating sharing; for example, it is explicitly cited as a rationale for institutional requirements and agreements such as the NIH data-sharing policy and the Bermuda Principles. But I do not think it can on its own explain the communist norm: we do not in general feel an obligation to do something just because it is socially beneficial, especially when there is substantial risk of personal harm—the optionality of charitable giving, lamented by Singer and Unger, being a notable example. Thus I will suppose in what follows that the communist ideal is its own thing, in need of its own special explanation.

communist norm, no governments or corporations or other funders with their own agendas—nothing except the priority rule, providing the background against which the advantages or otherwise of sharing are to be assessed. In such a world, what exchanges of information might a self-interested scientist willingly sign on to?

1.3 Contractual Sharing

1.3.1 One-Off Exchange

In a world where information is not freely shared, two scientists or two scientific research programs might nevertheless agree to exchange certain specified pieces of information for narrow, self-regarding reasons—that is, because each thinks they can improve their prospects by making the trade.

It is easy to see how agreements to share might be mutually beneficial when the parties are pursuing somewhat different goals. Such contracts may involve a "no trespassing" clause: the parties agree not to compete in the future, or even more explicitly, agree who has the right to publish and receive credit for which discoveries.

How, though, can a sharing agreement seem desirable to two research programs that are competing to make the same discovery, most particularly in the extreme case where they have precisely the same goals? You might reason as follows. Any trading of information must lead to one of two outcomes: either the trade will favor one side, increasing the chance that the favored program attains its goal before the other—that is, increasing the chances that one program will "scoop" the other—or it will be neutral, making a scoop no more likely than before. In the former case, the disfavored program will of course have good reason to eschew the trade. In the latter case, it seems that neither program will have reason to take the time and trouble to enter into an agreement. (Blumenthal et al. (2006) discuss the considerable work involved in preparing information for sharing.)

There are four reasons that a contract might nevertheless be feasible. First, one or both of the programs may have false beliefs about the effects of the trade, so that each thinks it has the advantage over the other. Second, in the case where the trade advantages neither party but has some net social benefit, altruistic desires may motivate both parties to expend the effort to make the exchange. Third, even if you are not in the least altruistic, some swaps might seem desirable although they directly benefit only your rival, just because your rival's winning the race to discovery, though it entails your failure, nevertheless produces some good for you as it does for all of science and society, in the form of new knowledge to consume. Fourth, both parties may gain an advantage as a result of the trade, not relative to one another, but relative to their competitors and, as it were, to nature.

I will put aside the explanation from ignorance, assuming for now that researchers can reliably estimate the benefits that will accrue from an information exchange. I also put aside the second and third explanations, from altruism and from the desire to know, for tactical reasons: if a communist contract is attractive even to scientists who entirely lack these motivations—as I eventually hope to show—then it must surely be all the more desirable to those who feel their pull.

To the fourth explanation, then. I will focus on a simple case in which two research programs are attempting to make a certain specific discovery or to reach some other specific goal, such as determining the structure of DNA, explaining superconductivity, or finding the genetic basis of sickle cell disease. (These are dramatic examples, but for the purpose of understanding sharing in a world governed by the priority rule, "goals" and "discoveries" include anything that would earn scientific credit—in other words, anything that a scientist would consider worthy of publication for its own sake.) Each program has its own particular methodology, or guiding theory, or expensive detector, and so there is no sense in which they could join forces to constitute a single coherent research program. They are by their very nature rivals, engaged in a race that only one of them can win.

There are three possible outcomes to such a race: one program makes the discovery first, the other program makes the discovery first, or neither makes the discovery. Suppose that a trade of information would decrease the third probability, and so increase the probability that some program makes the discovery. Suppose further that this increase to some extent benefits both of the rival programs. Then as a result of the trade, each will see an increase in its probability of making and receiving credit for the discovery. Thus, the trade is to the advantage of both programs—it is what I will call *mutually advantageous*.

Will the programs make the trade? If their only consideration is to maximize their expected credit earned, they will.[7] But in reality, they might reject such a trade if they perceive it to be unfair—if they think that their trading partner will benefit more (or considerably more) from the trade than they will. Or if they are able, they will negotiate a fairer deal. In short, they will arguably make only trades that are both mutually advantageous and fair. This concern for fairness is, I take it, a fundamental feature of human social psychology; it is to be accepted without further explanation as a fundamental constraint on contractual exchanges.

Many possible trades are mutually advantageous, increasing both trading partners' probabilities of discovery. But which of these will be regarded as fair?

Consider two proposals. First, a fair trade may be one that preserves the ratio of the programs' race-winning probabilities. Suppose, for example, that before the trade the probability that one program (call it the higher-power program) wins the race is 0.4, the probability that the other (lower-power) program wins the race is 0.2, and the probability that neither attains its goal is therefore 0.4. The higher-power program is, then, twice as likely to win the race as its lower-power competitor; the probability that the higher-power program wins, conditional on a discovery being made at all, is 2/3. Now suppose that the trade changes the probabilities to, respectively, 0.6, 0.3, and

[7] What if other, longer-term consequences are taken into account? You might refuse to make a trade that is mutually advantageous in the technical sense just defined if it is better for you that both you and your rival fail to make a discovery than that your rival succeeds and you fail—as when, for example, the winner of a discovery race not only receives credit for the discovery but also the lion's share of the funding for the next race. In these and similar circumstances, then, narrow self-interest militates against sufficiently unbalanced "mutually advantageous" trades. But I will assume that there is no need to take such considerations into account explicitly, as the work they do to motivate balanced trades is fully replicated by the concern for fairness described in the following paragraphs.

0.1. Then the higher-power program is still twice as likely to win as before; thus, the ratio-preservation criterion counts the trade as fair.

A second notion of fairness defines a trade as fair if it increases both programs' expected utility—their probability of winning the race multiplied by the benefits they gain from doing so—by the same amount. The benefits comprise the credit for making the discovery and all its indirect effects, such as increased funding for the next round of research and the thrill of victory (as well as perhaps the darker thrill of seeing your rivals fail). I assume that they are more or less the same for both programs—that the discoverers of the structure of DNA, for example, receive the same amount of credit, and feel the same level of satisfaction, whoever they happen to be.[8] Equal increases in expected utility, then, correspond to equal increases in each program's probability of winning. In the previous paragraph's scenario, for example, the trade is unfair in the present sense because it increases the higher-power program's probability of winning by 0.2 and the lower-power program's probability of winning by only 0.1—so increasing the higher-power program's expected utility by twice as much as the lower-power program's.

The second view of fairness is, I suggest, the one more likely to govern information trading and other arrangements in science. Or at least, it is the notion of fairness that we humans take to apply to contractual arrangements freely entered into: each party should put up goods of equal value. Thus a fair trade of information is one that increases each program's probability of winning by the same amount.

Three remarks. First, a program may under some circumstances be willing to make an unfair trade. Such decisions will depend on many factors, perhaps most of all the question whether it is in the programs' power to make a fairer trade. In some cases—if the exchangeable information comes in large, indivisible chunks—no fair trade may be possible, in which case the programs may enter into a mutually advantageous but unfair contract without the burden of resentment.

Second, the treatment of one-off contracts is easily extended to the case where there are other research programs engaged in the same race. A neutral trade may be mutually advantageous to the two traders if it improves their prospects, not relative to one another, but relative to a third program.

Third, an information exchange might be desirable not only if it increases the probability of the traders' making a discovery but also if it decreases the expected time taken to make the discovery: the sooner the discovery, the sooner the arrival of the endowed chair, the devoted graduate students, the corner office, and the big prize. Two programs will want to engage in a fair trade, then, if it decreases their expected waiting times to discovery, even if their race-winning probabilities remain the same.

To summarize, there are a range of circumstances under which, even before discovery and publication, contractual exchanges of information among programs will be beneficial to all parties to the exchange, and will be considered fair.

[8] I assume also, needless to say, that the utility of the scenario in which neither program makes the discovery is the same for both.

1.3.2 Open-Ended Exchange

Suppose that two research programs consider abandoning a regime of one-off exchanges for something more committing: an open-ended agreement to share information. Under what circumstances might they enter into such a contract?

Assume as in the previous section that the two programs are locked in a race to make the same discovery. In a one-off contract, what information is traded for what can be to a great degree specified in advance; the trade can typically be calibrated, then, so that both programs find it fair and mutually advantageous. The open-ended contract will of necessity be far less specific about what will be shared. What forms might it take?

I will consider two possible arrangements:

1. *Balanced sharing:* The programs undertake to share about the same amount of information with each other, and
2. *Total sharing:* The programs undertake to share any and all relevant information.

The first of these requires a little more structure; how to arrange things so that the equality of exchange is maintained? But I will not fuss about the details, as it will turn out that the second kind of agreement—the contractual equivalent of Merton's communism—is fairer and hence more agreeable to all parties.

How to compare the two proposals? Either, if implemented, will result in a certain number of discrete acts of sharing, concluding when the race to discovery is won (or peters out in failure). Consolidate these, so thinking of the consequence of the open-ended agreement as a single exchange in which each program gives to the other a package containing the sum total of the information disseminated over the duration of the race. An open-ended contract is desirable if this consolidated exchange is mutually advantageous and fair, using those terms just as I did above (but ignoring for simplicity's sake improvements in expected time to discovery):

Mutual advantage: Both programs' probabilities of winning the race go up at least slightly (at the expense of the probability that no program wins).

Fairness: Each program's probability of winning the race goes up by approximately the same amount.

So conceived, the open-ended contract is structurally identical to a one-off contract. The difference lies in the uncertainty surrounding the consequences of the agreement. Programs signing an open-ended contract will have to use some rather general principles to predict whether they are getting a fair and advantageous deal. We social scientists of science will have to do the same.

Let me begin by assuming that any exchange of information between research programs racing to make a discovery will diminish the probability that neither program makes the discovery, and that the complementary increase in some program's making a discovery benefits both programs at least a little.[9] It follows that any open-ended

[9] For circumstances under which, despite what I have said in the main text, one of the programs does not benefit from such an exchange, see note 11. See also note 7, although the concern discussed there is rendered moot by the fairness requirement.

exchange will be mutually advantageous to both programs. The remaining, and more difficult, question is which open-ended agreements will result in fair exchanges.

Break down into two components the probability that a research program wins the two-way race to discovery. The first component is the probability that our program would make the discovery in question, were it allowed to continue until it had exhausted its intellectual and physical resources. The second component is the probability that, if both our program and its rival do succeed in making the discovery, our program does so first. Call the probabilities respectively our program's *discovery probability* and its *race-clinching probability*. To continue the race metaphor, the discovery probability is the probability that the program crosses the finishing line; the race-clinching probability is the conditional probability that, if both it and its competitor both cross the line, it crosses before its competitor.

Two remarks. First, do not confuse the race-clinching probability with the unconditional probability of winning a race. This latter, *race-winning probability* is a function of the discovery and race-clinching probabilities of both programs (as explained in Section 1.6.1).

Second, in a scientific race, a program that is on track to make the discovery will normally abandon its research if it is scooped—if the discovery is made first by another program—since the priority rule awards no credit to runners-up. The event to which the discovery probability is attached, then, has a modal aspect, reflected in my language above: it is not the probability that the program actually makes the discovery (crosses the finish line), it is the probability that it would make the discovery (it would cross the line) if it were to stay in the race while remaining ignorant of its competitor's success.

The discovery and race-clinching probabilities may in principle vary somewhat independently of each other. There might be two strategies for realizing a goal, one slow and steady and the other risky but fast. A program adopting the first strategy will have a high probability of crossing the finishing line but will take a long time to do so; a program adopting the second strategy will have a low probability of crossing the finishing line, but if they get there at all they will get there quickly. Nevertheless, there are many factors that similarly impact both probabilities. The more numerous a program's scientific workers, or the more reliable its equipment, the more likely it is to reach its goal and to do so quickly. Similarly, programs that start out with some false assumptions are less likely to reach their destination, and even if they reach it, more likely to take their time, than programs that start out with mostly true assumptions (depending of course on the role of the assumptions). For these reasons, I will assume that a program's chances of success and its speed are correlated.

Suppose, then, using the same language as in the previous section, that each program has a certain degree of "power," and that a more powerful program is both more likely to make a discovery and more likely, if it makes a discovery, to make it relatively quickly.

Suppose also that if two programs exchange equal amounts of information, their respective discovery probabilities increase by the same amount.[10]

[10] There is reason to think that a higher-power program will benefit more from the same amount of information: it is in a better position to exploit that information. But there is also reason to think

In the light of these assumptions, let me consider in turn the two possible open-ended information-sharing agreements laid out above: the balanced sharing agreement, according to which programs share equal amounts of information, and the total sharing agreement, according to which programs share all relevant information.

If programs share equal amounts of information then their discovery probabilities increase by equal amounts. You might think that this means that their probabilities of winning the discovery race increase by equal amounts too, in which case the balanced sharing rule results in trades that are fair for everyone. But this is not quite correct.

To see why, consider an especially simple model of a discovery race. The model assigns to each program a power between zero and one, and stipulates the following consequences of power:

1. A program's discovery probability is equal to its power.
2. The higher-power program has a race-clinching probability that is greater than one-half; since race-clinching probabilities sum to one, the lower-power program has a race-clinching probability of less than one-half.

These assignments can be thought of as notionally effecting the following protocol for determining the winner of a discovery race:

1. Each program has a probability equal to its power of being placed into the pool of would-be winners.
2. From the pool of would-be winners, the actual winner is determined by the race-clinching probabilities.

Being placed into the pool represents, then, the outcome to which a program's discovery probability is attached: its making the discovery eventually under its own steam. If both programs make it into the pool, then being chosen from the pool as the winner represents the outcome to which a program's race-clinching probability is attached: its being the first to make the discovery, given that both programs make it (or rather, would make it) at all.

In what follows, I examine the consequences of an especially basic version of the simple model in which the higher-power program's race-clinching probability is equal to one, implying a zero race-clinching probability for the lower-power program. When both programs make it into the would-be winners' pool, then, the higher-power program invariably wins the race.

Suppose that two programs follow the balanced sharing agreement, exchanging equal amounts of information. By assumption, they gain equally as far as their discovery probabilities—their chances of admission to the would-be winners pool—are concerned. But what is the impact on what ultimately matters, their probabilities of winning the race? If the value of admission to the pool changed commensurately for

that a lower-power program will benefit more: it is less likely to have already in its possession an alternative solution to the problem solved by the information (a specific case of the phenomenon of diminishing marginal returns). Perhaps these roughly balance out, yielding something like the same information/same increase principle proposed in the main text.

both programs, or remained the same, then this equal gain in the chances of admission would bring about equal increments in the race-winning probabilities, and hence the expected utilities, of both programs. But the value of admission to the pool changes unequally: it stays the same for the higher-power program, but it decreases for the lower-power program.

Why? The value of admission to the pool does not change for the higher-power program because the higher-power program's race-clinching probability is equal to one: when it makes it into the pool, it always wins. The lower-power program wins, by contrast, only when it makes it into the pool and the higher-power program does not. Its expected value of admission to the pool decreases, then, as the higher-power program's probability of admission increases.

In short: as the discovery probabilities of the two programs increase, their probabilities of admission to the would-be winners' pool increase commensurately, but whereas the value of admission to the pool remains the same for the higher-power program, it decreases for the lower-power program. Consequently, equal increments of discovery probability do not benefit the higher- and lower-power programs equally; the higher-power program does better. A trade of equal amounts of information between the two programs is for this reason unfair; a fair trade would require the higher-power program to put up more information than the lower-power program.[11]

The same is true for more realistic parameterizations of the simple model in which the race-clinching probabilities lie between zero and one (Section 1.6.2). What matters is that, as programs share information, there are more likely to be multiple would-be winners. A high-power program, with its higher race-clinching probability, does relatively better in such situations than a low-power program, and so will tend to benefit more from an equal exchange of information. To share equally is not to share equitably; on the contrary, fair sharing requires more powerful programs to give more—an observation that I will call the *Marxian precept*. "From each according to its ability ..."

Fair-minded researchers will, as a consequence of the Marxian precept, refuse to enter an open-ended contract that requires balanced sharing. How will they regard the other contract suggested above, the total sharing agreement that requires programs to exchange all information in their possession? Suppose that the amount of information possessed by a program is proportional to its power. Then the effect of a total sharing contract is that lower-power programs will share less information and higher-power programs more. But that is just to implement the Marxian precept, and so to do what is needed—qualitatively speaking—to effect an open-ended exchange that is not only mutually advantageous but fair.

1.3.3 Toward Universal Exchange

If an open-ended agreement to share any and all information can make sense between two research programs of unequal power in direct competition, why not among many

[11] When the discovery probabilities are already high, an equal exchange may even result in an absolute decrease in the lower-power program's race-winning probability. Exchanges of information that increase discovery probabilities are not always advantageous for both programs, then. In the main text I quietly pass over this possibility, though it provides a further consideration against balanced sharing.

research programs, some going head to head, some partially competing, and many related only in passing? Might not such an arrangement seem to most or all researchers likely to result in exchanges that on balance improve everyone's prospects while implementing the Marxian precept? If so, they should be willing to sign a social contract mandating total sharing—the Hobbesian equivalent of science's communist norm.

To get from the fairness of a two-way open-ended exchange to the fairness of a universal social contract, various bridges must be crossed. Does it make a difference, for example, that many programs participating in the great information exchange are no longer competing directly? Not obviously, since the less the competition, the less the worry that by giving away information, you are giving your rivals an unfair advantage. A more serious question concerns the enforcement of the contract—though the answer to this question depends rather a lot on what you aim to do with the Hobbesian vindication.

Concerning these issues, I have almost nothing further to say in this chapter. They must eventually be tackled, but it is far more important to examine the weak points in the argument for total sharing in the two-program case.

Let me review the assumptions driving that argument:

1. Higher-power research programs have higher discovery and race-clinching probabilities.
2. An information exchange increases a program's discovery probability by an amount proportional to the amount of information exchanged.
3. The simple model yields a realistic estimate of the changes in research programs' expected utilities effected by various trades.
4. The amount of information that a research program has to share is roughly proportional to its power.

The first three assumptions establish the Marxian precept, the qualitative claim that fairness requires higher-power programs to share more than lower-power programs, and the fourth takes you from the Marxian precept to the approximate fairness of total sharing.

The latter three assumptions are rather simplistic, but it is not ridiculous to suppose that (2) and (4) hold on average over a large number of information transfers, given a suitable notion of "amount of information," or that they are as good as anything available for the purposes of deliberating about whether to sign an open-ended contract. The third, however, looks to be badly wrong.

In the simple model, the two programs' race-clinching probabilities—their probabilities of winning the race, given that both make it into the pool of would-be winners—remain the same after the information exchange. (I set the probabilities equal to zero and one in the exposition in the main text; in the generalization in Section 1.6.2 they may take on any values provided that the higher-power probability is greater than the lower-power probability, but still they are held constant through the exchange.) However, you might reasonably surmise that the higher-power program's race-clinching probability would decrease and the lower-power program's probability would increase as a result of the exchange, on the grounds that, if the two programs' discovery probabilities increase by the same amount, the ratio of the probabilities decreases, and thus the potential (in some sense) of the lower-power program increases relative to that of the higher-power program. Suppose, for example, that the discovery probabilities of the two programs go from 0.4 and 0.2 respectively to 0.6 and 0.4. Then the lower-power program goes from having half the potential (in some sense) of the higher-power program to having two-thirds its potential. Why not think that it enjoys also a commensurate increase in relative speed?

This would work to nullify at least partly the effect so important in the argument for the Marxian precept, in which entry to the pool of would-be winners becomes less valuable for the lower-power program as the discovery probabilities of the two programs increase, so would in part undo the rationale for total sharing.

To address this worry, and to better understand the structure of scientific races more generally, let me construct a more sophisticated model of discovery and information exchange. It will turn out that, though the simple model is itself too simple, its key property—the preservation of race-clinching probabilities necessary to secure the argument for the Marxian precept—is surprisingly robust.

1.4 A Waiting-Time Model of Discovery

1.4.1 The Discovery Density

The subject matter of the new waiting-time model is the same sort of idealized race to make a single discovery that was treated in previous sections. To review the essential assumptions about such a race: Each research program in the race has as its only goal to make the designated discovery. If and only if it does so before any other program, it "wins" the race. There are two ways the program might fail to win the race: it might because of faulty assumptions or some other mishap not be in a position to make the discovery at all, that is, it might fail to cross the finishing line (a possibility quantified by the program's discovery probability), or it might for systematic reasons or from sheer bad luck fail to make the discovery soon enough, crossing the finishing line behind some other program with the same goal (a possibility quantified by the program's race-clinching probability).

The waiting-time model associates with each research program a probability distribution representing the chance of its making the discovery in any given time period, which I call the program's *discovery distribution*. The discovery distribution can be represented by a probability density function $f(t)$, defined so that the probability that the program makes its discovery between any two given points of time is equal to the area under the density between the corresponding values of t (Figure 1.1).[12]

The probability that the program makes its discovery before time T, then, is the area under the density function to the left of T. The probability that the program, if allowed to go on indefinitely, at some point makes its discovery—what I have been calling the

[12] I assume that the density satisfies the constraints obviously implied by this interpretation: it is zero for values of t less than zero and the area under the density is not greater than one. Because the area may be, and usually is, less than one, the discovery density is not strictly speaking a probability density. But in every other respect it behaves like one.

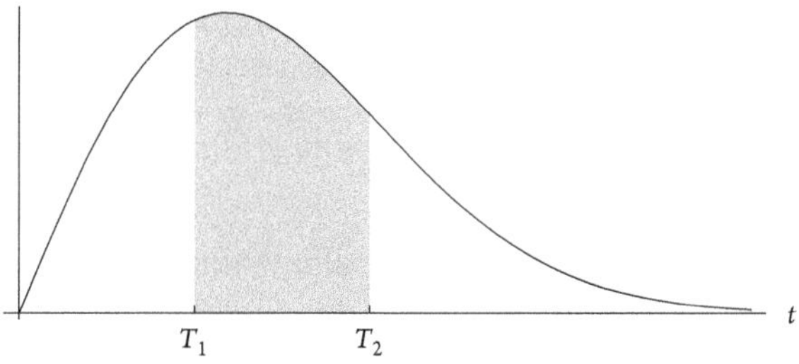

Figure 1.1 The discovery distribution: the probability that the program makes its sought-after discovery between T_1 and T_2 is the area under the density between those two points (shaded).

program's discovery probability—is the total area under the density. This probability is typically, I will suppose, less than one.

The model itself does not assume any particular interpretation of the discovery distribution, but since the probabilities are supposed to explain what exchanges of information scientists regard as mutually advantageous or fair, I will suppose that the distribution represents scientists' subjective probabilities at a point in time where a trade is being considered. For simplicity's sake, I will assume that all scientists engaged in a race see the situation in the same way, and so have the same subjective probabilities at any point in time. Nothing nearly so strong is necessary, however; scientists' conclusions about sharing will turn out to be much the same whatever their probabilities. There is no need to assume that the scientists' probabilities remain fixed over time, since the question examined by this chapter is of what trades seem mutually advantageous or fair at any given moment.

As remarked above, the discovery probabilities themselves are inherently subjunctive. The area under the density is the probability that a program *would* make its discovery, if allowed to go on until it reaches its natural end. When in a priority race one program makes the discovery, the other program, having nothing to gain from further research, typically gives up. In that case, we will never know whether the event of the losing program's making the discovery, if allowed to go on indefinitely, "occurred." But in the model, the probability of this event is perfectly well defined, and the supposition that scientists make use of such probabilities in their deliberations is eminently reasonable, I think, even though they only occasionally have the chance to calibrate them by observing the frequencies of the events in question.

Suppose that several research programs are competing to make the same discovery. Then, on the assumption that their discovery probabilities are stochastically independent (see note 14), their discovery densities determine the values of the race-clinching and race-winning probabilities. A program's race-winning probability, for example, is the sum, for every possible time, of the probability that the program makes the discovery at that time multiplied by the probabilities that each of the other programs fails to make the discovery by that time. (Such quantities are better expressed mathematically, for which see the mathematical development in Section 1.6.3.2 and Section 1.6 more generally.)

The discovery densities of the programs engaged in a particular race, then, comprise a simple but complete stochastic model of the race. It is this model that I will use to investigate the connection between boosts in discovery probabilities and race-clinching probabilities.

1.4.2 The Effect of Inflation

The argument for the fairness of total sharing turned on the argument for the Marxian precept, that is, the qualitative claim that higher-power research programs ought, if sharing is to be fair, to give away more information than lower-power programs. That argument proceeded by examining the fairness of an equal exchange of information, showing that higher-power programs benefit more from such exchanges than low-power programs.

Two assumptions did the work in this demonstration of the unfairness of equal sharing between unequally powerful programs:

1. The race-clinching probability is higher for higher-power programs.
2. When programs trade equal amounts of information, so boosting their discovery probabilities, their race-clinching probabilities do not change.

It is the second of these assumptions that I questioned above. It is true by fiat in the simple model, but that appeared to be a tendentious idealization. I now show that appearances are misleading; the assumption is quite broadly valid.

Suppose, then, that two programs are engaged in a discovery race, one of higher power than the other. They exchange equal amounts of information. The result of the exchange, I assumed above, is an equal boost in discovery probability, in absolute rather than relative terms; for example, both programs' discovery probabilities increase by 0.1. The question: do such boosts alter race-clinching probabilities?

The answer is that it depends on the details. Let me begin by examining what is perhaps the simplest form of boost, what I will call *uniform inflation* (or inflation for short). A uniform inflation increases a program's discovery probability by uniformly "blowing up" its discovery density, or in other words, by multiplying the density by some number greater than one, as shown in Figure 1.2.

What effect does inflation resulting in equal boosts have on race-clinching probabilities? None whatsoever. Indeed, the size of the boosts is immaterial. Inflate or deflate

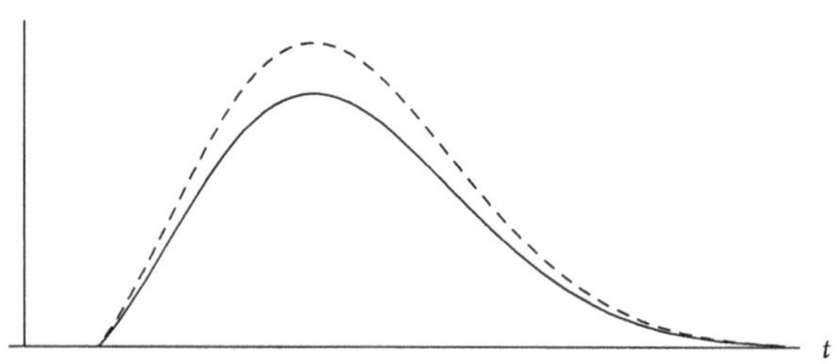

Figure 1.2 Uniform inflation. The original density is plotted with a solid line, the inflated density with a dashed line.

two competing programs' discovery densities by any amounts—inflate one and deflate the other, if you like—and there will be no change to the race-clinching probabilities. A proof is given in Section 1.6.3.3; the reason in words is that the clinching probabilities are conditional on both programs making their discoveries, which means that the magnitudes of the discovery probabilities are rendered irrelevant. All that matters is the way that one program's discovery probability is distributed in time relative to the other's, something that inflation leaves unchanged.[13]

Provided that information exchange results only in uniform inflation, the Marxian precept holds true. But why suppose that the effect of an information exchange is confined to inflation? Could what is learned from other research programs not move the discovery density's mass along the axis of time? The next section will be devoted to examining such operations.

1.4.3 Advancement and Compression

How might new information, or any other improvement in a research program's resources, affect its discovery density if not by inflation? Two simple possibilities are (Figure 1.3):

Advancement: As a result of the new information, the density is shifted to the left.
Compression: As a result of the new information, the density is "squeezed," into a smaller range, while preserving the total area under the density.

Whereas inflation affects a program's probability of discovery, advancement and compression affect the expected speed of discovery. To say the same thing more formally, advancement and compression, unlike inflation, decrease a program's expected waiting time to discovery (conditional on a discovery's being made), while leaving the discovery probability itself untouched.

Suppose that, as a result of their exchanging equal quantities of information, two programs' discovery densities are advanced by the same amount, that is, shifted equal distances to the left. What is the effect on the race-clinching probabilities? Obviously, none: the two programs are in the same relative position as they were before. A formal argument is given in Section 1.6.3.4.

[13] If a program's race-clinching probability does not depend on the area under its discovery density—on its overall probability of making a discovery—but only on the way that whatever probability it has is distributed relative to the probability of the other program, then why suppose, as I did in the argument for the Marxian precept, that the higher-power program has a higher race-clinching probability? I gave some answers in Section 1.3.2, but the discussion of inflation provides good reason to reconsider these carefully.

One answer in particular withstands, I think, reevaluation in the light of the proof. Of two competing research programs, the higher-power program—the program with the higher probability of making its discovery—will tend to attract more scientists (Kitcher 1990; Strevens 2003). With more scientists at work, the higher-power program is likely to move through the steps on the road to discovery relatively more quickly than the lower-power program. Its expected time to discovery conditional on its making a discovery at all will tend to be less than that of the lower-power program, then, and so its race-clinching probability will tend to be higher.

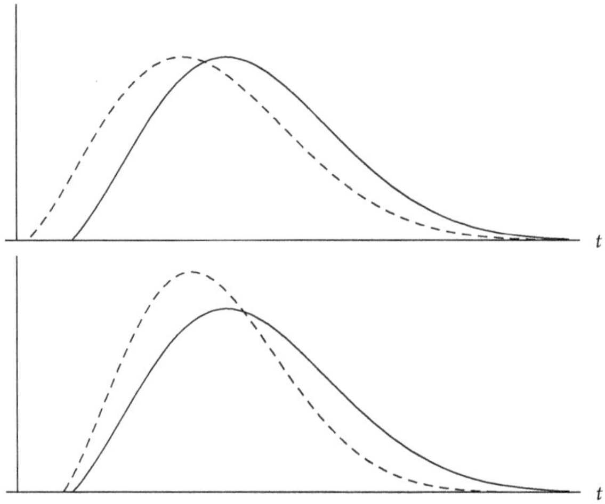

Figure 1.3 Advanced (top) and compressed (bottom) discovery densities. The original densities are plotted with solid lines; the transformed (i.e., advanced or compressed) densities are plotted with dashed lines.

The same is true for compression: equal compressions of two programs will leave their race-clinching probabilities unchanged (Section 1.6.3.5). (By contrast with the case of inflation, note, it is crucial for the preservation of the race-clinching probabilities that the degree of advancement or compression is the same for both densities.)

It can be concluded not only that individual inflations, equal advancements, and equal compressions preserve race-clinching probabilities, but that any combination of these operations also does the same. If an equal exchange of information results in both an equal compression and an inflation of the programs' discovery densities, for example, the race-clinching probabilities are unaffected and so the exchange benefits the higher-power program more—vindicating the Marxian precept. The precept holds true, then, for a wide range of exchange-induced transformations.

How might that range be expanded?

1.4.4 A Multiple-Stage Model of Discovery

Uniform inflation, advancement, and compression are all what you might call global operations on a discovery density, affecting all regions of the density equally. Might the effect of some information exchanges be more regional? What if a research program's path to discovery has multiple steps, and a certain body of information bears only on one of the steps?

To investigate this possibility, I introduce an enhancement of the waiting-time model that I call the multiple-stage model of discovery; I will use it to show that race-clinching probabilities are remarkably resistant to change even when it is supposed that new information has a purely regional impact.

Consider a research program with two stages. The first stage must be completed successfully in order for the second stage to begin. If the second stage is then completed successfully, the discovery is made.

The two stages are assigned completion densities; these are the analog of discovery densities, giving the probability that a stage is completed within any given period of time (relative to the time that work on the stage begins).

Assume that the two stages are stochastically independent (although of course the second stage will be attempted only if the first stage is successfully completed). Then the program's discovery density is determined by the stages' completion densities as explained in Section 1.6.4.1.

Now suppose that an information exchange boosts the completion probability of the first stage only, by uniformly inflating its completion density. What is the effect on the discovery density for the program as a whole? It is uniform inflation by the same amount (Section 1.6.4.2). Causally, the effect on the program is regional, but probabilistically, it is global, of a sort that we already know has no effect on race-clinching probabilities.

The same is true of advancement: advancing the completion density for the first stage by a certain amount advances the discovery density of the whole by the same amount (Section 1.6.4.3). And also for compression: compressing the first stage's completion density by a certain amount compresses the discovery density by roughly the square root of that amount (Section 1.6.4.4).

The result may be fully generalized: it holds for research programs with more than two stages, and for transformations applied not only to the first but to any such stage (Section 1.6.4.5).

Consequently, inflating, advancing, or compressing the completion densities for single stages of two competing multiple-stage research programs yields the same result, relatively speaking, as applying the operations to the discovery density as a whole. Any inflation of single stages, and any equal advancement or compression of single stages, will therefore leave race-clinching probabilities unchanged.

1.5 Hobbesian Sharing

Self-interested scientists, I have argued, ought to see the value of participating in a (properly policed) regime of total information sharing. That is the Hobbesian vindication of the communist norm. The vindication relies on a number of assumptions, some prima facie plausible and one—the preservation of race-clinching probabilities through information exchanges—prima facie implausible but nevertheless correct.

Two pressing worries remain. First, I assumed back in Section 1.3.1 that scientists were accurate in their estimates of research programs' prospects—that is, accurate in their estimates of the programs' discovery and race-clinching probabilities, and by extension, of their discovery densities. You might also see the need to assume agreement among scientists on how to measure information, and so on what constitutes an exchange of equal amounts of information, and agreement on the effects of information exchange on the densities. Are scientists really so well informed? Or at least, so well coordinated in their judgments?

The Hobbesian vindication of communism in fact requires very little agreement on these things. Scientists might have quite divergent views about the prospects of this or that program, or about what goes into a quantum of information. Or they might have no particular opinion at all. What my argument for a universal commitment to the Marxian precept requires is that each scientist is able to proceed from whatever assumptions they have to the conclusion that total sharing is considerably more fair than balanced sharing, and considerably more advantageous than no sharing. For that, they need only agree to the premises used in my argument above. I assumed no particular form for the discovery densities, and no particular scheme for measuring information. Rather, I called upon certain considerably more abstract propositions: that discovery probability is correlated with race-clinching probability; that exchanges will result in some combination of inflation, advancement, and compression; that an information exchange that increases the participating programs' discovery probabilities by about the same amount will compress their discovery densities (if at all) by about the same amount; and so on. That is all the scientists themselves need to believe in order to appreciate the merits of total sharing.

It is arguably much easier, then, to reach the conclusion that total sharing is a good idea than to negotiate a particular one-off information exchange—for which agreement on the details of densities and information most definitely is required.

A second and perhaps even more important worry is that the qualitative Marxian precept falls short of establishing the quantitative conclusion that a total sharing contract is totally fair. Balanced sharing may unfairly favor higher-power programs, but it does not follow that the optimal amount of sharing will be achieved by total sharing rather than some other loosely Marxian scheme, that is, some other scheme that asks higher-power programs to share more.

To these concerns the Hobbesian should respond as follows. There are only a handful of universally binding, open-ended sharing contracts that are practically feasible, in the sense that they can be implemented at a reasonable cost given the available resources. "Everyone should share everything" is certainly feasible, as is "No one is obliged to share a thing," which being the state of nature implements itself. Balanced sharing—"everyone should share equally"—is more complicated to put into practice, but arguably still within reach.

It is among the feasible alternatives that the scientist of practical necessity chooses. They can do nothing, in effect choosing the state of nature—no socially organized sharing—or they can choose balanced or total sharing. It is difficult to think of other plausible candidates. (Share half of what you have? Insist on receiving twice as much as you dispense?) The question, then, is not whether one of these is optimal, all practical difficulties aside, but whether one is clearly better than the others, with practical difficulties (including enforcement) figured into the cost.

Though I happily concede that there is little reason to think total sharing to be the fairest policy among all logical possibilities, I have shown it to be clearly more attractive than the other practical possibilities. On the one hand, the exchanges resulting from a total sharing contract will be fairer than those resulting from a balanced sharing contract. On the other hand, they will be far more advantageous (given proper enforcement) to individual programs than the state of nature, and they seem to be fair enough—no scientist has reason to expect that they will be systematically short-changed under a total sharing contract. So Hobbesian scientists should sign the damn thing.

A Hobbesian vindication of the communist norm is a model-theoretic tool rather than an end in itself. How to put it to work? I have two purposes in mind.

The first is to provide fortification for communist practice. Giving away information is never rational from a narrow, self-interested point of view. Various aspects of modern-day science accentuate the possibility of loss: the profit motive for scientists producing commercially useful research; the fight for tenure-line positions and then for tenure for scientists in the university system; and the superstar economy in general. But society has a powerful interest in total scientific sharing. It may be helpful to point out to scientists, then, that in the long term, and with adequate encouragement or enforcement, they will be even in the narrowest sense better off under a continuation of the communist regime.

Observe that the argument for the contract is genuinely Hobbesian, not merely Rawlsian. Total sharing need not be seen through a veil of ignorance to seem beneficial to the deliberator. Even scientists who know quite well that they are in a relatively high-power research program, and who care for nothing but their own welfare, will see the case for the contract—as even the best-armed strongman, Hobbes thought, even the king himself, should see that his life is excruciatingly vulnerable in a war of all against all.

The second application of the Hobbesian vindication is less useful but more ambitious: correctly deployed, it will surely explain the origins of the communist norm. In what way that explanation might proceed is a matter I cannot explore here. There are many questions to answer, and these require the knowledge and skills not only of philosophers but also of historians, sociologists, psychologists, and anthropologists among others. How was the line from a contractual to a moral obligation crossed—or was there never a contractual phase? Did scientists, calculating or intuiting the virtues of a total sharing contract, at some point recent or distant deliberately write it into their moral code? Did the norm arise by some process of cultural evolution inside the petri dish of western European science (in which case it is reality's probability distributions, not scientists', to which the assumptions made in this chapter will have to apply)? Or does it go back further? Perhaps science has inherited from humanity an age-old social norm governing the sharing of a broad class of information, exemplified by but not exhausted by scientific knowledge. Perhaps, along with egalitarianism, communism of a limited sort is buried deep in the human mind. Perhaps we have always believed, to give the communist norm its liberal expression, that information wants to be free.

1.6 Appendix: Mathematical Development

1.6.1 Race-Winning and Race-Clinching Probabilities

Consider a two-way race to make a single discovery between research programs **F** and **G**. Let F be the event that **F** would make the discovery at some point, if allowed to go on indefinitely, and G the corresponding event for **G**. Then FG is the event that both programs would make the discovery, and hence the event of there being what you might call a "live race." There are two possible outcomes for a live race: either **F** makes the discovery first and so wins the race, or **G** does. Call the first outcome $F^{*}G$

and the second outcome FG^*; the asterisk, then, indicates the winner. If **F** makes (or rather would at some point make) its discovery, then one of three mutually exclusive events must occur: F^*G, FG^*, or $F\neg G$, the last being the case where **G** fails to make the discovery at all. (I ignore dead heats.) Thus:

$$P(F) = P(F\neg G) + P(F^*G) + P(FG^*).$$

The probability $P(F^*)$ that **F** wins the race, live or not, and so gains credit for the discovery, is what I have called **F**'s *race-winning probability*. It is the sum of the probabilities of two outcomes corresponding to the two ways that a program can win a discovery race: first, that **F** makes the discovery and **G** does not; second, that **F** and **G** both make the discovery—the race is "live"—but **F** does so first. In symbols:

$$\begin{aligned} P(F^*) &= P(F\neg G) + P(F^*G) \\ &= P(F) - P(FG^*). \end{aligned}$$

The probability that **F** wins a given live race is what I have called **F**'s *race-clinching probability*; it is the conditional probability $P(F^*G \mid FG)$. Since F^*G occurs only if FG occurs, the race-clinching probability of **F**, which I will write as $C(F)$, is

$$C(F) = P(F^*G)/P(FG).$$

1.6.2 The Marxian Precept

The mathematical component of the argument for the Marxian precept, the demonstration that exchanges of equal amounts of information disproportionately benefit the program with the higher race-clinching probability (assumed to be the higher-power program), can be formalized as follows.

The benefit bestowed upon a program by an information exchange is directly proportional (I assume in the main text) to the resulting increase in its probability of winning the discovery race. What must be shown, then, is that equal boosts to discovery probabilities result in unequal boosts to race-winning probabilities, with the higher-power program benefiting more.

The probability that **F** wins its race is (from Section 1.6.1)

$$\begin{aligned} P(F^*) &= P(F) - P(FG^*) \\ &= P(F) - P(FG)C(G). \end{aligned}$$

Likewise, the probability that **G** wins is $P(G) - P(FG)C(F)$. Suppose that the discovery probabilities $P(F)$ and $P(G)$ are boosted by the same amount. Then the win probabilities go up by that amount, less the right-hand terms in the expressions above—for **F**, the term $P(FG)C(G)$. Because $P(FG)$ increases, this "penalty" term increases for both programs (if both race-clinching probabilities are non-zero). But it increases more for the program with the lower race-clinching probability, by assumption the lower-power program.

What if the distance between the race-clinching probabilities is narrowed by the exchange? This may compensate for the unequally weighted increase in $P(FG)$, resulting in no net change in the right-hand-side "penalties," in which case an equal information exchange would be fair.

If there are more than two programs competing in the discovery race the mathematics gets tedious, with many terms representing the many different possible outcomes, but I believe that the same result holds for qualitatively the same reasons.

1.6.3 The Waiting-Time Model

1.6.3.1 *Discovery Densities*

Consider a two-way race between research programs **F** and **G** with discovery densities $f(t)$ and $g(t)$ respectively.

The probability that **F** makes the designated discovery before time T is by definition the area under the density function to the left of T:

$$\int_0^T f(t)\,dt.$$

The probability that **F** would make the discovery at some point or other, that is, **F**'s discovery probability, is:

$$P(F) = \int_0^\infty f(t)\,dt.$$

1.6.3.2 *Race-Clinching Probabilities*

Recall from Section 1.6.1 that **F**'s race-clinching probability is defined to be $P(F^*G)/P(FG)$, where FG is the event that both **F** and **G** make the discovery and F^*G is the event that both programs make the discovery and **F** does so first.

Both $P(F^*G)$ and $P(FG)$ are fixed by the discovery densities $f(t)$ and $g(t)$, if it is assumed that F and G are stochastically independent—an assumption I make throughout this development, rationalized in the accompanying footnote.[14] The probability of FG:

$$\begin{aligned} P(FG) &= P(F)P(G) \\ &= \int_0^\infty f(t)\,dt \int_0^\infty g(t)\,dt. \end{aligned}$$

And the probability of F^*G:

$$P(F^*G) = \int_0^\infty f(t)G(t,\infty)\,dt, \tag{1.1}$$

where $G(a,b)$ is the definite integral of $g(t)$ between a and b.

[14] The independence assumption can be only approximately correct. Suppose that the two research programs are built on similar presuppositions; the success of one program in making a discovery would then suggest that the presuppositions are true and so that the other is more likely to succeed. Equally, if they disagree on a presupposition, their discovery probabilities will for the same sort of reason be negatively correlated. Another source of correlation is programs' exchanging "information" that has some chance of turning out to be misleading. As best I can see, these failures of independence will not tend to pull race-clinching probabilities systematically in one direction or another: race-clinching probabilities are quotients of the form $P(F^*G)/P(FG)$, and the correlations discussed in this note will (I surmise) tend to have much the same effect on both numerator and denominator, either increasing both or decreasing both in at least rough proportion and so making little difference overall.

To investigate the effect of information exchange on race-clinching probabilities, then, is to investigate its effect on the ratio of these two expressions.

1.6.3.3 *Inflation*

A uniform inflation of a discovery density multiplies it by some number greater than one, transforming it from $f(t)$ to $jf(t)$.

Suppose that two programs' discovery densities are uniformly inflated, not necessarily by the same factor. What is the effect on their race-clinching probabilities? The answer is: none.

Proof: Suppose that **F**'s density is inflated by a factor of j and **G**'s by k. Let $\hat{P}(\cdot)$, $\hat{f}(t)$, and $\hat{g}(t)$ be the new probability distribution and the new discovery densities respectively—that is, the distribution and densities that result from inflation. (Thus, $\hat{f}(t) = jf(t)$ and $\hat{g}(t) = kg(t)$.) The aim is to show that, assuming independence of the discovery distributions, **F**'s post-inflation race-clinching probability $\hat{P}(F^*G \mid FG)$ is equal to its pre-inflation race-clinching probability $P(F^*G \mid FG)$.

Because integration is linear, the definite integral functions will receive the same boost as the densities, so that for example the post-inflation definite integral $\hat{G}(T,\infty)$ is the pre-inflation definite integral boosted by k, or in symbols $\hat{G}(T,\infty) = kG(T,\infty)$. Linearity also underlies the proof as a whole:

$$\begin{aligned}
\hat{P}(F^*G \mid FG) &= \frac{\hat{P}(F^*G)}{\hat{P}(F)\hat{P}(G)} \\
&= \frac{\int_0^\infty \hat{f}(t)\hat{G}(t,\infty)\,dt}{\int_0^\infty \hat{f}(t)\,dt \int_0^\infty \hat{g}(t)\,dt} \\
&= \frac{\int_0^\infty jf(t)kG(t,\infty)\,dt}{\int_0^\infty jf(t)\,dt \int_0^\infty kg(t)\,dt} \\
&= \frac{jk\int_0^\infty f(t)G(t,\infty)\,dt}{jk\int_0^\infty f(t)\,dt \int_0^\infty g(t)\,dt} \\
&= \frac{jkP(F^*G)}{jkP(F)P(G)} \\
&= P(F^*G \mid FG).
\end{aligned}$$

Inflation by any mix of inflating (or deflating) factors therefore leaves race-clinching probabilities unchanged.

1.6.3.4 *Advancement*

Suppose two research programs receive an equal advancement, transforming their discovery densities from $f(t)$ and $g(t)$ to $f(t+j)$ and $g(t+j)$ respectively. What is the effect on the race-clinching probabilities?

Consider **F**'s race-clinching probability $P(F^*G)/P(FG)$. Because the advancement has no effect on the areas under either density, there is no effect on $P(F)$ or $P(G)$, hence (given independence) on the denominator $P(FG)$.

Intuitively, there is also no effect on the numerator $P(F^*G)$. For a formal proof, integrate equation (1.1) by substitution with $u = t + j$, then note that for advancement to make sense, the original discovery densities for both programs must be zero for values of t between 0 and j.

The race-clinching probability is therefore unchanged.

1.6.3.5 *Compression*

In mathematical terms, $f(t)$ is compressed by transforming it to $jf(jt)$ (for $j > 1$; if j is less than one, you have rarefaction rather than compression). Suppose that two programs' discovery densities are compressed or rarified by the same factor j. Then the race-clinching probabilities are left unchanged.

Proof: Since compression, like advancement, preserves the total area under the discovery densities, what must be shown is that the probability of F^*G is unchanged by the transformation. As before, let $\hat{P}(\cdot), \hat{f}(t)$, and $\hat{g}(t)$ be the transformed probability distribution and the transformed discovery densities respectively. Thus

$$\hat{P}(F^*G) = \int_0^\infty \hat{f}(t)\hat{G}(t,\infty)\,dt.$$

Consider $\hat{G}(t,\infty)$ first. Let $u = jt$ and integrate by substitution with $du = j\,dt$:

$$\begin{aligned}\hat{G}(T,\infty) &= \int_T^\infty \hat{g}(t)\,dt \\ &= \int_T^\infty jg(jt)\,dt \\ &= \int_{jT}^\infty g(u)\,du \\ &= G(jT,\infty).\end{aligned}$$

Returning to the probability of F^*G, integrate by substitution with the same variable:

$$\begin{aligned}\hat{P}(F^*G) &= \int_0^\infty \hat{f}(t)\hat{G}(t,\infty)\,dt \\ &= \int_0^\infty jf(jt)G(jt,\infty)\,dt \\ &= \int_0^\infty f(u)G(u,\infty)\,du \\ &= P(F^*G).\end{aligned}$$

1.6.3.6 *Races with Many Programs*

The proofs above are straightforwardly generalized to races in which there are three or more programs competing to make the same discovery. To see this, observe that for three programs **F**, **G**, and **H**:

$$P(F^*GH) = \int_0^\infty f(t)G(t,\infty)H(t,\infty)\,dt$$

and so on. The rest is left as an exercise to the reader.

1.6.4 The Multiple-Stage Model

1.6.4.1 *Two-Stage Model*

Let the completion densities for a two-stage research program be $f_1(t)$ and $f_2(t)$. The first density gives the probability that the first stage is completed at any particular time; when it is completed, work begins on the second stage, completion of which constitutes discovery. The time taken to discovery is of course the sum of the time taken to complete each stage.

Given stochastic independence of the stages, the program's discovery density is:[15]

$$f(T) = \int_0^T f_1(t)f_2(T-t)\,dt.$$

1.6.4.2 *Inflation of a Single Stage*

Suppose that the completion probability of a two-stage model's first stage is uniformly inflated by a factor j, yielding a new completion density $jf_1(t)$. Then the new discovery density for the program as a whole is:

$$\begin{aligned}\hat{f}(T) &= \int_0^T jf_1(t)f_2(T-t)\,dt\\ &= jf(T).\end{aligned}$$

The effect of a boost to the first stage is therefore a uniform inflation of the entire discovery density by the same factor.

1.6.4.3 *Advancement of a Single Stage*

Suppose that the completion density of a two-stage model's first stage is advanced by a factor j, yielding a new completion density $f_1(t+j)$. Then the new discovery density for the program as a whole is:

$$\hat{f}(T) = \int_0^T f_1(t+j)f_2(T-t)\,dt.$$

Define a new variable $u = t + j$ and apply integration by substitution, obtaining

$$\hat{f}(T) = \int_j^{T+j} f_1(u)f_2(T+j-u)\,du.$$

Now consider the effect of advancing the original discovery density $f(T)$ by a factor of j, that is, consider $f(T+j)$:

$$f(T+j) = \int_0^{T+j} f_1(t)f_2(T+j-t)\,dt.$$

[15] Because the completion densities are equal to zero for all negative values of t, this definite integral is equal to the definite integral over the entire range of t, that is, from $-\infty$ to ∞. It is therefore the convolution of $f_1(t)$ and $f_2(t)$, a fact I will put to work in Section 1.6.4.4.

You will see that

$$\hat{f}(T) = f(T+j) - \int_0^j f_1(t)f_2(T+j-t)\,dt.$$

Advancing $f_1(t)$ by j makes sense only if the density is zero for values of t less then j (otherwise, you are moving some probability mass into t's negative realm); thus, the right-hand term is zero and so the effect of advancing the first stage of a two-stage project is to advance the discovery density as a whole by the same amount.

1.6.4.4 *Compression of a Single Stage*

Suppose that the completion probability of a two-stage model's first stage is compressed by a factor j, yielding a new discovery density $\hat{f}(T)$ for the program as a whole. Ideally I would show that the compression of the first stage is equivalent to a compression of the whole, that is, that there exists some factor k such that

$$\hat{f}(T) = kf(kT).$$

This does not hold true generally, but it is approximately true for an interesting and large subclass of cases, provided that the compression factor j is fairly close to 1.

Let $\phi(s)$ be the characteristic function of the original discovery density and $\phi_1(s)$ and $\phi_2(s)$ be the characteristic functions of the first and second stages' original completion densities respectively. Then because the discovery density is the convolution of the completion densities (see note 15),

$$\phi(s) = \phi_1(s)\phi_2(s).$$

When the first-stage completion density $f_1(t)$ is compressed by a factor j, it becomes $jf_1(jt)$. What is the effect on the characteristic function? Let $\hat{f}_1(t)$ denote the transformed function and $\hat{\phi}_1(s)$ its characteristic function. Then

$$\begin{aligned}
\hat{\phi}_1(s) &= \int_{-\infty}^{\infty} e^{ist}\hat{f}_1(t)\,dt && \text{(definition of characteristic function)}\\
&= \int_{-\infty}^{\infty} e^{ist} jf_1(jt)\,dt\\
&= \int_{-\infty}^{\infty} e^{isu/j} f_1(u)\,du && \text{(letting } u = jt\text{)}\\
&= \phi_1(s/j).
\end{aligned}$$

Thus, the characteristic function of the discovery density as a whole becomes

$$\hat{\phi}(s) = \phi_1(s/j)\phi_2(s).$$

We want this to equal the characteristic function of $kf(kT)$ for some k, which is, using the same reasoning, $\phi(s/k)$, and so by the convolution theorem is equal to

$$\phi_1(s/k)\phi_2(s/k).$$

The goal, then, is to show that there is a k such that

$$\phi_1(s/j)\phi_2(s) = \phi_1(s/k)\phi_2(s/k).$$

It will be especially nice if the correct value for k depends only on j.

Now we approximate. Suppose that j is sufficiently close to one that the characteristic functions of the completion densities are approximately linear over the intervals $[s/j, s]$, for all relevant values of s. Restrict the search for k to the range $j \geq k \geq 1$, so that s/k will fall into such an interval.

Over such an interval, it follows from approximate linearity that there exist constants a, b, c, and d such that, for any s in the interval,

$$\phi_1(s) \approx as + c \quad \text{and} \quad \phi_2(s) \approx bs + d.$$

Then

$$\begin{gathered}\phi_1(s/j)\phi_2(s) \approx \phi_1(s/k)\phi_2(s/k) \Rightarrow \\ (a\frac{s}{j} + c)(bs + d) \approx (a\frac{s}{k} + c)(b\frac{s}{k} + d) \Rightarrow \\ ab\frac{s^2}{j} + ad\frac{s}{j} + bcs + cd \approx ab\frac{s^2}{k^2} + ad\frac{s}{k} + bc\frac{s}{k} + cd.\end{gathered}$$

One way for this approximate equality to hold is for the following two approximate equalities to hold:

$$\frac{ab}{j} \approx \frac{ab}{k^2} \qquad \text{(constraint A)}$$

$$ad\frac{s}{j} + bcs \approx ad\frac{s}{k} + bc\frac{s}{k} \qquad \text{(constraint B)}$$

Constraint A implies that

$$k \approx \sqrt{j}.$$

I will show that setting k to the square root of j is also sufficient for constraint B to hold approximately. Since the value of k does not depend on the interval, it is good for the whole function, which is what we need.

Solving constraint B's approximate equality (or rather, the corresponding exact equation) for k:

$$\begin{aligned} ad\frac{s}{j} + bcs &= ad\frac{s}{k} + bc\frac{s}{k} \Rightarrow \\ \frac{ad}{j} + bc &= \frac{ad + bc}{k} \Rightarrow \\ k &= \frac{j(ad + bc)}{ad + jbc}. \end{aligned}$$

Since a, b, c, and d are constants, there exists some z such that $ad = zbc$. The above expression for k can then be simplified:

$$\begin{aligned} \frac{j(ad + bc)}{ad + jbc} &= \frac{j(zbc + bc)}{zbc + jbc} \\ &= \frac{j(z + 1)}{z + j}. \end{aligned} \qquad (1.2)$$

For values of z with a large magnitude, equation (1.2) is approximately equal to j; for values close to zero, it is approximately equal to 1. For intermediate magnitudes it takes values from just less than 1 to j.[16] Since j is close to 1, any value for k between 1 and j will satisfy constraint B at least approximately; the choice of $\sqrt{j}$ in particular, being about halfway between 1 and j, will do nicely. That choice also, of course, satisfies constraint A.

Note that, when z is approximately equal to 1, equation (1.2) is approximately equal to the square root of j, so that constraint B is satisfied almost exactly.

Proof: taking z as equal to 1 and squaring equation (1.2) gives you

$$\frac{4j^2}{(j+1)^2} = \frac{4j^2}{j(j+2+1/j)}$$
$$\approx \frac{4j^2}{4j} \approx j$$

since for j close to 1, $j+1/j$ is almost exactly equal to 2. When is z close to 1, then? That is, when is ad roughly equal to bc? Suppose that the completion densities for the first and second stages have roughly the same form, modulo an inflationary factor y. Then

$$f_1(t) = yf_2(t) \Rightarrow \phi_1(t) = y\phi_2(t)$$
$$\Rightarrow a = yb \quad \text{and} \quad c = yd$$

for any interval over which the characteristic function is approximately linear. Thus,

$$ad = yb\frac{c}{y} = bc$$

and so $z = 1$.

Taking stock, we did not get an exact formula for k that depends only on j. We did get something close: setting k equal to the square root of j will give us approximately what was wanted, and when the completion densities for the two stages have roughly the same form (modulo an inflationary factor), almost exactly what was wanted—all provided that the compression factor j is close to 1.

It remains to be asked to what degree the approximations made along the way—in particular the assumption of the linearity in the small of the characteristic functions and the fudging in the satisfaction of constraint B—diminish the interest of the result. As it happens, for many relevant probability densities, such as the gamma distributions often used to represent waiting-time probabilities, the match between the discovery density with a compressed first stage and the compressed discovery density (using the square root approximation) is very close, and almost exact when the functions have the same form.

[16] Exception: equation (1.2) is badly behaved when z is in the vicinity of -1. On intervals where z is close to -1, however, the terms in the expressions for $\phi_1(s/j)\phi_2(s)$ and $\phi_1(s/k)\phi_2(s/k)$ with the coefficients ad and bc will roughly cancel out, so that for the purpose of finding a good value of k for these intervals, constraint B can be ignored. In that case, setting k to $\sqrt{j}$ so as to satisfy constraint A will be sufficient for what is wanted.

1.6.4.5 Generalization

The results above are demonstrated only for a certain subclass of multi-stage research programs, namely, those that have two stages and in which the transformation in question—inflation, advancement, or compression—is applied to the first of those stages. The results are, however, easily and fully generalized.

First, transforming the first stage of a two-stage program has the same effect as transforming the first stage of a multi-stage program, since any number of stages after the first can be represented by a single completion density representing the probability distribution over the time taken for all stages after the first to complete. Simply interpret the second completion density in the proofs above as representing such a distribution.

Second, transforming the second or later stage of a multi-stage program has the same effect as transforming the first stage, since the completion densities for each of the stages are multiplied together in the expression for the discovery density, and multiplication commutes.

Acknowledgments

Thanks to Thomas Boyer-Kassem and to audiences at the PSA 2008, the conference on Collective Knowledge and Epistemic Trust at the Alfried Krupp Wissenschaftskolleg, Greifswald, the Colombian Conference on Logic, Epistemology, and the Philosophy of Science at the Universidad de los Andes, Bogotá, Carnegie Mellon, and the University of Barcelona.

References

Arzberger, P., P. Schroeder, A. Beaulieu, G. Bowker, K. Casey, L. Laaksonen, D. Moorman, P. Uhlir, and P. Wouters. (2004). Promoting access to public research data for scientific, economic, and social development. *Data Science Journal* 3:135–152.

Blumenthal, D., E. G. Campbell, M. Gokhale, R. Yucel, B. Clarridge, S. Hilgartner, and N. A. Holtzman. (2006). Data withholding in genetics and the other life sciences: Prevalences and predictors. *Academic Medicine* 81:136–145.

Boyer, T. (2014). Is a bird in the hand worth two in the bush? Or, whether scientists should publish intermediate results. *Synthese* 191:17–35.

Collins, H. M. (1974). The TEA set: Tacit knowledge and scientific networks. *Science Studies* 4:165–185.

Contreras, J. L. (2011). Bermuda's legacy: Policy, patents, and the design of the genome commons. *Minnesota Journal of Law, Science & Technology* 12:61–125.

Dasgupta, P. and P. A. David. (1987). Information disclosure and the economics of science and technology. In G. Feiwel (ed.), *Arrow and the Ascent of Modern Economic Theory*, 519–542. New York University Press, New York.

———. (1994). Toward a new economics of science. *Research Policy* 23:487–521.

Heesen, R. (2016). *To Expedite the Flow of Knowledge*. Ph.D. thesis, Carnegie Mellon University.

Hereditary Disease Foundation. (2014). Hereditary disease foundation FAQ. Accessed online 1/21/14 at http://www.hdfoundation.org/faq.php.

Jasonoff, S. (2006). Transparency in public science: Purposes, reasons, limits. *Law and Contemporary Problems* 69:21–45.

Kitcher, P. (1990). The division of cognitive labor. *Journal of Philosophy* 87:5–21.

Louis, K. S., L. M. Jones, and E. G. Campbell. (2002). Sharing in science. *American Scientist* 90:304–307.

Macfarlane, B. and M. Cheng. (2008). Communism, universalism and disinterestedness: Re-examining contemporary support among academics for Merton's scientific norms. *Journal of Academic Ethics* 6:67–78.

Merton, R. K. (1942). The normative structure of science. *Journal of Legal and Political Sociology* 1:115–126. Originally titled "Science and Technology in a Democratic Order." Page references are to the reprint in Merton (1973)

______. (1957). Priorities in scientific discovery. *American Sociological Review* 22:635–659.

______. (1973). *The Sociology of Science*. University of Chicago Press, Chicago.

National Institutes of Health. (2003). Final NIH statement on sharing research data. Posted at http://grants.nih.gov/grants/guide/notice-files/NOT-OD-03-032.html.

Resnik, D. B. (2006). Openness versus secrecy in scientific research. *Episteme* 2:135–147.

Savage, C. J. and A. J. Vickers. (2009). Empirical study of data sharing by authors publishing in PLoS journals. *PLoS ONE* 4(9):1–3.

Strevens, M. (2003). The role of the priority rule in science. *Journal of Philosophy* 100:55–79.

Tenopir, C., S. Allard, K. Douglass, A. U. Aydinoglu, L. Wu, E. Read, M. Manoff, and M. Frame. (2011). Data sharing by scientists: Practices and perceptions. *PLoS ONE* 6(6):1–21.

Wade, N. (1981). *The Nobel Duel: Two Scientists' 21-Year Race to Win the World's Most Coveted Research Prize*. Anchor Doubleday, New York.

Watson, J. D. (1968). *The Double Helix*. Atheneum, New York.

2

Publish Late, Publish Rarely!: Network Density and Group Performance in Scientific Communication

STAFFAN ANGERE AND ERIK J. OLSSON

2.1 Introduction

Group inquiry is an essential aspect of scientific activity, as we know it. It takes place not only in research groups inquiring into some part of nature or culture, but also in various scientific committees and boards deciding what scientific projects to fund, whom to appoint for an academic position, etc. Outside science, group inquiry, and deliberation play a prominent role in juries in court, in political committees, in the corporate boardroom, and in many other places. Whereas in the infancy of scientific investigation much activity arguably took place at the individual level, this is not how inquiry is conducted in many, if not most, areas of modern science. And of course, the possibility of instant electronic communication has increased group activity and interaction dramatically.

But is communication actually beneficial in these cases? Does it actually make for better conclusions or decisions? The question may seem unexpected, even imbecile. Of course it is good to communicate, one is inclined to exclaim. And indeed, this gut reaction is perfectly appropriate in many cases in science and elsewhere. Here is one reason for this: Group inquiry typically focuses one or more research questions to be solved by collective means. By a complex question we mean a question that cannot be answered without being further divided into subordinate questions. Thus, the question whether John committed the robbery is more complex than the question whether he had a motive. To be answerable, the former needs to be decomposed into more specific questions, including questions about his motive. Complex questions abound in science. For instance, the question whether there is man-made global warming is a complex question par excellence, presupposing, for instance, an answer to the question whether a higher level of CO_2 causes an increase in global temperature. Now, many complex questions are naturally, or even necessarily, treated by means of a division

Boyer-Kassem, Mayo-Wilson, & Weisberg. *Scientific Collaboration and Collective Knowledge*.

of labor where different research groups work on different subproblems requiring different competences. At some point, these groups need to pool their results, i.e., communicate their findings among the members, or the overarching question cannot be answered. So in this case it can hardly be denied that communication is beneficial, indeed essential, just as common sense insists it should be. In the following we will focus instead on (relatively) simple questions that do not obviously call for such division of labor.

Suppose that we wish to maximize group performance in solving a common problem. Suppose, moreover, that the group members are determined, and that we can play with the communication structure of the group, that is, we can decide who is to communicate with whom. If so, how should we "hook people up"? Common sense dictates that the more group members can communicate with each other, the more competent the group will become. If so, then a complete graph, where there is a communication link from everyone to everyone else, is the way to go.

Unfortunately, studies in various disciplines—including social psychology and economics—persistently undermine, or problematize, the common-sense answer. These studies have found a lot of communication links, i.e., high *network density*, to be detrimental to group performance in a wide range of cases. In section 2, we review four such studies. In the remainder of the chapter, we explore the issue within the framework of our own Bayesian model of communication, implemented in the simulation environment Laputa. We introduce this model in sections 3 and 4. In section 5, we describe the setting within which our simulation experiments are carried out. As we report in section 6, one experiment shows that the "repugnant conclusion" that network density is generally detrimental to group performance can be replicated in our framework. This leads us to asking whether increasing the quality of information in the network can make network density more attractive. We study, in section 7, the effect of imposing two such conditions, which effectively disallow inquirers from *recycling* already posted information under two different interpretations of what recycling means, and we observe that network density can be positively correlated with group performance, construed as average closeness to the truth (in a sense to be defined), if these conditions are imposed. We also observe that there is a trade-off to be made between group performance and group polarization.

2.2 Background

Economists Bala and Goyal (1998) studied decision-making in a group context using mathematical modeling and analysis. The inquirers in the model could choose between two actions without knowing which action is optimal. They could make use of information regarding the payoff of their own previous action, and the payoffs of their neighbors in the network. As an example the authors take a case of medical inquirers working on a particular disease. There are two possible states of the world:

$$\varphi_1 = \text{The old method is better.}$$

$$\varphi_2 = \text{The new method is better.}$$

There are, correspondingly, two actions inquirers can take:

$$A_1 = \text{Pursuing the old method.}$$

$$A_2 = \text{Pursuing the new method.}$$

Because the old method is assumed to be well understood, pursuing it will not result in any new information about the state of the world. Unbeknownst to our inquirers, the new method is in fact better. Inquirers can see the results of some of the other inquirers. This gives rise to a graph structure where the nodes represent inquirers and the links communicational connections. Inquirers update their subjective probabilities of φ_2 based on their own results and the result of the other inquirers they can see by means of Bayesian conditionalization. A population of inquirers has finished learning just in case one of the following conditions has been met: all inquirers take action A_1 or all inquirers believe that they are in φ_2 (with probability greater than 0.9999). A population of inquirers has learned successfully just in case all inquirers believe that they are in φ_2 (with probability greater than 0.9999). Otherwise they have "learned unsuccessfully". Bala and Goyal now show that, if inquirers are arranged on a line, all inquirers will learn successfully, i.e., believe truthfully that the new method is better. Surprisingly, however, if there is, in addition, a small "royal family" of inquirers communicating with all the others, inquirers sometimes end up learning unsuccessfully, i.e., pursuing the worse method. Hence the royal family network, which is more "connected," is actually less reliable, collectively, than the linear network. Bala and Goyal conclude that "more informational links can increase the chances of a society getting locked into a sub-optimal action" (p. 609).

Bala and Goyal's study raises a number of questions. What is it about the group with a "royal family" that reduces group performance? Is it the particular communicational graph structure? Is it the fact that the graph is more dense, by which we mean that it has a higher number of edges per node? What role does the sheer size of the group play (in cases of finite groups)? Some answers can be found in Zollman (2007), which is a detailed study of Bala and Goyal's model for finite groups of inquirers using computer simulation. Zollman compared networks with a cycle or wheel structure with the complete graph regarding probability of successful learning and learning speed. He also looked at the effect of group size on these matters. Zollman went on to study the general relationship between communication graph density and group performance for different group sizes. He found that sparser networks are generally more reliable than denser ones. However, learning speed increases with density. In other words, a denser network learns faster but less reliably.

Mason, Jones, and Goldstone (2008) studied the propagation of "innovations" in networked groups in laboratory experiments with real participants. Participants had their computers arranged in a virtual network structure and made repeated numerical guesses for which they received scores. The scores were also made available to their neighbors in the network. The task was for each participant to maximize his or her score over 15 rounds. The networks were compared on speed of discovery and convergence to the optimal solution (within a small margin of error) for a number of different score (payoff) functions with different shapes where the score functions used were unknown to the participants. The percentage of participants who guessed the number with the highest global score as a function of the number of "rounds" was studied. Mason et al. found that denser networks are generally faster but less reliable than sparser networks, such as a "small world" network: "The advantage of the small-world to the fully connected network is akin to a novel group-based form of the 'less is more' effect reported in individual decision making literature" (2008, p. 430). A suggestive result concerns a case of an easy-to-find local maximum but a difficult-to-find global maximum. The fully connected network converged quickly on the local maximum and stayed there. The lattice structure, composed of "small research groups," was eventually more reliable in finding the global maximum. The study of Mason et al. indicates that the network structure may influence the learning speed as well as the reliability of a group in ways that depend on the structure of the problem space, i.e., on things like how many locally or globally optimal solutions there are, and how easy it is to find those solutions.

Lazer and Friedman (2007), surveying research in this area, reached similar conclusions: "Remarkably, the highest performing network in the long run was the slowest, most error prone, and had the longest average path length. More generally, our results highlight that given a long time horizon, efficient communication can actually be counterproductive to systemic performance, whereas given a short time horizon, more communication generally contributes to better outcomes. Our results thus suggest a potential dark side to the rush to connectedness in today's world: it will limit our pathways to developing new and better ways of doing things, even as the demand for new solutions is increasing."

A reasonable conjecture is that the negative conclusions in the literature, as for as the epistemic value of network density is concerned, are partly due to the effect of network "spamming": as the density of the network increases, it becomes more sensitive to low-quality information. An interesting question which is not dealt with systematically in these studies, or any other study we know of, is whether the conclusions carry over to cases where people are more discriminate as to when to post information in the network, for instance by refraining from posting things whose truth value they are not quite sure of. We will study this issue in the context of the model implemented in the software Laputa, which we have developed for the purpose of studying epistemic properties of networked communication, drawing on the influential approach to social epistemology in Goldman (1999).[1]

2.3 A Model of Scientific Inquiry and Communication

We define a *research network N* as a group of people or other agents (including some organizations) engaging in communicative and investigative practices. Examples include university departments, peer-review boards, circles of friends, or even entire societies. As a limiting case, the entirety of humanity during a period can be seen as such a network. The "during a period" clause is necessary here: we are taking the

[1] See, for instance, Masterton and Olsson (2013), Olsson (2011, 2013), Olsson and Vallinder (2013), Vallinder and Olsson (2013, 2014). Laputa can be downloaded at http://www.luiq.lu.se/portfolio-item/laputa/.

network, i.e., the communicative and investigative practices themselves, to be fairly stable. We also assume that the participants in the network remain fixed. Members of a research network are referred to as *inquirers*.

We assume that there is a single empirical proposition p whose truth value the inquirers are trying to ascertain. As the model we are using is Bayesian, we take each inquirer α, at each time t, to have a certain *credence* $C^t_\alpha(p)$ in p. This value represents each inquirer's current knowledge about the state of p. But we also have to model the way that the participants receive new information. There are two fundamentally different inlets for this: inquiry and communication.

Inquiry can here be taken to include any kind of method of altering a credence function which does not base itself on information given by anyone else in the network; in physicists' terms, it corresponds to *external* forces. Some paradigmatic examples of inquiry are observation, experiment, and perhaps taking advice from persons outside N. In order for our model to be applicable, the opinions of inquirers in N must have fairly little effect on such "external" persons' opinions, however. An example would be a modern philosopher reading Immanuel Kant, who does not himself need to be taken as a part of our social network, since his opinions cannot be affected by anything we say or do.

Not all inquirers' approaches to inquiry are the same, and they tend to vary both in their degree of activity and their effectiveness. Let $S^t_{t\alpha}(p)$ be the proposition "α's inquiry gives the result that p at time t," let $S^t_{t\alpha}(\neg p)$ be the proposition "α's inquiry gives the result that not-p at t," and let $S^t_{t\alpha} \equiv_{df.} S^t_{t\alpha}(p) \vee S^t_{t\alpha}(\neg p)$ be the proposition that α's inquiry gives *some* result at t. We represent the participants' properties qua inquirers by two probabilities: the chance $P(S^t_{t\alpha})$ that, at moment t, α receives a result from her inquiries, and the chance $P(S^t_{t\alpha}(p)|S^t_{t\alpha}, p)$ that, when such a result is obtained, it is the right one.[2] $P(S^t_{t\alpha})$ will be referred to as α's *inquiry chance*, and $P(S^t_{t\alpha}(p)|S^t_{t\alpha}, p)$ as her *inquiry accuracy*.

It is important to keep in mind that these, unlike $C^t_\alpha(p)$, are *objective chances*. They can be interpreted statistically, or perhaps in terms of propensities. In our implementation of the model they will be fed into a pseudorandom number generator to determine what happens during a simulation.

An inquirer without interest in p would generally have a low value of $P(S^t_{t\alpha})$, while one very interested in p, but engaging in inquiry using faulty methods, would have a high value of $P(S^t_{t\alpha})$, but an inquiry accuracy close to 0.5, or even below that. In the latter case, the results of her inquiry would actually be negatively correlated with the truth. As a simplification, we will assume α's inquiry chance and accuracy to be constant over time, so we will usually write them without the time index t.

Just as inquiry represents the flow of information *into* the network, communication deals with how this information is disseminated *within* the network. While, as a first approximation, we may be interested only in whether or not α can receive information from β, we generally need to go deeper. Thus, we take the network N to

[2] This is only half the story: $P(S^t_{t\alpha}(p)|S^t_{t\alpha}, p)$ gives the chance of a true report when p is the case, and does not say anything about reliability when $\neg p$ holds. To simplify matters, we assume that these probabilities are equal, i.e., that the chance that an agent's inquiry gives the correct result does not depend on whether p is true or false.

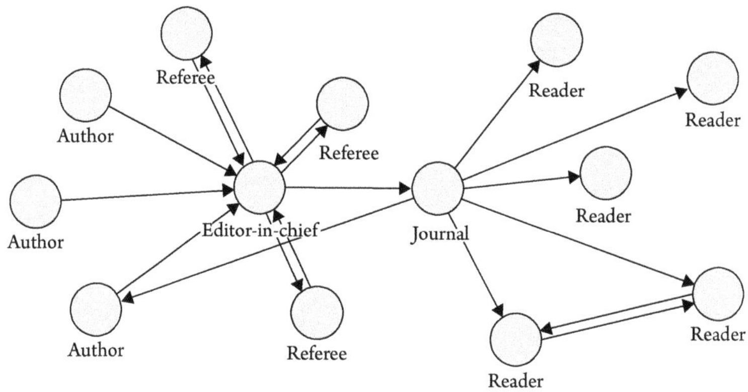

Figure 2.1 A research network

include a set of what we will call *links*, each corresponding to a communication channel. Such channels can be as direct as conversation, or mediated like a blog or an instant messaging system. Generally, however, we may want to say that if the "mediator" is able to choose which messages to transmit, it may be more apt to be represented as an inquirer instead of a link, even if it engages in no inquiry itself (this is one case where we may want to set $P(S^t_{t\alpha}) = 0$). A scientific journal could be an example of this, as in the network depicted in figure 2.1.

The links from authors to the editor describe the submission process, the links between the editor and the referees the refereeing process, and the links from the journal to the other inquirers their individual practices of reading that journal.

Like inquirers, links have an internal structure. While links do not have anything corresponding to inquiry accuracy, since the messages they transmit are chosen by the senders and thus not subject to random variation, they have degrees of activity. Two inquirers who seldom talk have weak links between them, while researchers who sit in the same room may be assumed to have stronger links. Analogously to the inquiry notation we define

$$S^t_{\beta\alpha}(p) \equiv_{df.} \beta \text{ says that } p \text{ to } \alpha \text{ at } t,$$

$$S^t_{\beta\alpha}(\neg p) \equiv_{df.} \beta \text{ says that not} - p \text{ to } \alpha \text{ at } t,$$

$$S^t_{\beta\alpha} \equiv df. \beta \text{ says either that } p \text{ or that not} - p \text{ to } \alpha \text{ at } t.$$

The strength of a link $\beta\alpha$, called its *communication chance*, is representable as a probability $P(S^t_{\beta\alpha})$, being the chance that β communicates that p or that not-p to α, at any given moment t, given that she is certain enough that p (or not-p) is indeed the case.

Given that β does communicate with α, what does she say? Prima facie, it seems that β somehow would indicate how strongly she believes that p, i.e., she would communicate her credence. But this credence is not in general available to her. As a *subjective degree of belief*, it is a product of betting behavior, and it takes experiments to determine it rather than introspection. We may of course represent β's beliefs about her own degree of belief as credences as well, but this only pushes the problem further back: somehow, we will still need to determine exactly what to say, given a credence distribution over possible degrees of belief.

Instead, we will adopt a simplified model, which still has enough complexity to be able to represent the important aspects of the situation. We assume that unless she is out to mislead, β will say p if she believes p, and not-p if she believes not-p. But belief is not an all-or-nothing matter in the Bayesian tradition, so how sure must she be of p to say it? This question is answered by a property of the link $\beta\alpha$ that we will call its *certainty threshold*: a value $\theta_{\beta\alpha}$ between 0.5 and 1 such that p can be communicated over $\beta\alpha$ only if $C^t_\alpha(p) > \theta_{\beta\alpha}$, and $\neg p$ can be communicated over $\beta\alpha$ only if $C^t_\alpha(p) < 1 - \theta_{\beta\alpha}$. This combines with the communication chance so that the actual probability of communicating becomes $P(S^t_{\beta\alpha})$ if $\theta_{\beta\alpha}$ is passed, and 0 otherwise.

Are there communicative practices not representable as links of this kind? There is the possibility that β is a systematic liar, but for our current application we will disregard that. We have also already mentioned the scientific journal, which may be more reasonable to model as an inquirer. But what of mass-broadcasting systems, like a TV station, or the above-mentioned blog? A link, as we have defined it, has a single source and a single recipient. However, nothing stops us from representing a broadcasting system by a set of links—one for each viewer. Since $P(S^t_{\beta\alpha})$ is the probability that α actually hears what β says, rather than just the probability that β says something, even a broadcasting system ought to have different links for different viewers.

2.4 Trusting Oneself and Others

We have described how the participants in a research network N engage in inquiry and communicate, but we have as yet said nothing about how they react to the results of these practices. It seems that, in general, hearing that p from someone, or receiving a result indicating p from inquiry, should influence an inquirer's credence in p. Fortunately, Bayesianism has a universal answer to the question of how this should be done: belief update proceeds through conditionalization.

But conditionalizing on p whenever one hears that p is not reasonable. Straight conditionalization really works only for infallible sources, and in general an inquirer only takes messages proclaiming p as indications of its truth, and not as conclusive verification. One solution is to use Jeffrey conditionalization instead of regular conditionalization (Jeffrey 1983). This is more of a promissory note than a solution as such, since we still need to decide on a degree of belief to set p to when one hears that p. Furthermore, inquirers can receive several messages at the same time, and some of these may even contradict one another, so what we need is a framework that allows us to handle such cases.

One way forward is as follows. To start with, it seems admissible for α to treat both her own inquiry and the things said to her in roughly the same way: as indications of whether or not p is the case. We refer to α's inquiry ι and the other inquirers $\beta, \gamma, \ldots$ who can talk to her as her *sources*. Each of these may have different connections to the truth, and we can represent these by the probability that a source gives the right answer. More specifically, we define *source σ's reliability for α* as

$$r_{\sigma\alpha} =_{df.} P\left(S_{\sigma\alpha}\left(p\right) \mid S_{\sigma\alpha}, p\right) = P\left(S_{\sigma\alpha}\left(\neg p\right) \mid S_{\sigma\alpha}, \neg p\right).$$

This definition presupposes that the probability that any source gives the answer p if p is the case is equal to the probability that it gives the answer not-p if not-p is the case. This *source symmetry* simplifies our calculations, although it can be relaxed if we encounter cases where it does not provide a reasonable approximation.[3] As with other invocations of P, it is worth remarking that $r_{\sigma\alpha}$ is concerned with objective chances rather than credences, although inquirers will have credences in propositions such as $a < r_{\sigma\alpha} < b$ which are *about* such chances. In fact, P will only appear in such credences, so in this case we do not need to worry about the interpretation of objective chance at all.

It is quickly checked that the reliability of α's inquiry is identical to her inquiry accuracy. For other sources, it is an abstraction based on those sources' performances as indications of truth. In general, an inquirer has no direct access to this value, but this does not stop her from forming beliefs about it. Since the number of possible values for the chance $r_{\sigma\alpha}$ is infinite, we need to represent α's credence as a density function instead of a regular probability distribution. Thus, for each inquirer α, each source σ, and each time t, we define a function $\tau^t_{\sigma\alpha} : [0,1] \to \mathrm{R}$ which we call α's *trust function* for σ at t, such that

$$C^t_\alpha\left(a < r_{\sigma\alpha} < b\right) = \int_a^b \tau^t_{\sigma\alpha}\left(x\right) dx$$

for a, b in $[0,1]$. This function is uniquely defined up to a set of measure 0, according to the Radon-Nikodym theorem. $\tau^t_{\sigma\alpha}(x)$ gives the credence density at x and we can obtain the actual credence that α has in propositions about the reliability of her sources by integrating this function. We will also frequently use the expression $1 - \tau^t_{\sigma\alpha}(x)$, which represents α's credence density for propositions about σ *not* being reliable, and we will refer to this function as $\overline{\tau^t_{\sigma\alpha}}(x)$. More generally, we will use the bar x" as a shorthand for the expression $1 - x$.

Now, it is obvious that an inquirer's credences about chances should influence her credences about the outcomes of these chances. The way this should be done is generally known under a name David Lewis gave to it: the principal principle.[4] This says that if α knows that the chance that an event e will happen is r, then her credence

[3] There is another way to define reliability, namely, as the probability that p is the case, given that p is reported. The two concepts are, however, interdefinable through the equation

$$P^t_\alpha\left(p \middle| S_{\alpha\sigma}\left(p\right)\right) = P^t_\alpha\left(S_{\alpha\sigma}\left(p\right) \middle| S_{\alpha\sigma}, p\right) \frac{P^t_\alpha\left(S_{\alpha\sigma}\left(p\right) \middle| S_{\alpha\sigma},\right)}{P^t_\alpha\left(p \middle| S_{\alpha\sigma},\right)}.$$

Because the definition used here is simpler for our purposes, and because (as a referee pointed out) $P^t_\alpha(p|S_{\alpha\sigma}(p))$ may be difficult to interpret given that p is simply true or false, we have adopted this version. The difference corresponds to Goldman's (1986) distinction between *reliability* and *power*.

[4] Lewis (1980). An earlier version is given by Miller (1966). Although there has lately been a large debate on when these principles are admissible, we will only be using two specific instances of the general case. None of these appear to give rise to any problems.

in e should be exactly r. Applied to our case, this means that the following dual principle (PP) must hold:

$$a < C^t_\alpha \left(S^t_{\sigma\alpha}(p) \mid S^t_{\sigma\alpha}, p, a < r_{\sigma\alpha} < b\right) < b$$
$$a < C^t_\alpha \left(S^t_{\sigma\alpha}(\neg p) \mid S^t_{\sigma\alpha}, \neg p, a < r_{\sigma\alpha} < b\right) < b$$

Hence, α's credence in σ giving the report p, given that she knows that the source gives any report at all, that p is actually the case, and that σ's reliability is between a and b, should itself be between a and b, and similarly for $\neg p$. Since $r_{\sigma\alpha}$ can take a continuum of values, we have framed the principle in terms of intervals rather than specific values. A quantity such as $C^t_\alpha(r_{\sigma\alpha} = x)$, on the other hand, would be 0 for almost all values of x.[5]

We also have use for an independence postulate. While not strictly necessary, such a postulate will simplify calculations and modeling considerably. The independence assumption we use here will be referred to as *communication independence* (*CI*):

$$C^t_\alpha(S^t_{\sigma\alpha}, p, a < r_{\sigma\alpha} < b) = C^t_\alpha(S^t_{\sigma\alpha}) \times C^t_\alpha(p) \times C^t_\alpha(a < r_{\sigma\alpha} < b).$$

Communication independence implies that whether σ says anything is independent of whether or not p actually is true, as well as of what reliability σ has. This is true in the current model, since we have assumed that a source's reliability for reporting that p, given that p is the case, is the same as its reliability for reporting that $\neg p$ is the case, given that $\neg p$.

From *PP* and *CI* we can derive the following expression for α's credence in σ's reliability (see the appendix for the actual derivation):

$$C^t_\alpha\left(S^t_{\sigma\alpha}(p) \mid p\right) = C^t_\alpha(S^t_{\sigma\alpha}) \times E\left[\tau^t_{\sigma\alpha}\right]. \tag{2.1}$$

Here, $E\left[\tau^t_{\sigma\alpha}\right]$ is the expectation of the trust function $\tau^t_{\sigma\alpha}$. We refer to this as the link $\sigma\alpha$'s *expected trust*. Using equation (2.1), an application of Bayes' theorem together with the law of total probability gives us

$$C^t_\alpha\left(p \mid S^t_{\sigma\alpha}(p)\right) = \frac{C^t_\alpha(p) \times E\left[\tau^t_{\sigma\alpha}\right]}{C^t_\alpha(p) \times E\left[\tau^t_{\sigma\alpha}\right] + C^t_\alpha(\neg p) \times E\left[\overline{\tau^t_{\sigma\alpha}}\right]}.$$

$$C^t_\alpha\left(p \mid S^t_{\sigma\alpha}(\neg p)\right) = \frac{C^t_\alpha(p) \times E\left[\overline{\tau^t_{\sigma\alpha}}\right]}{C^t_\alpha(p) \times E\left[\overline{\tau^t_{\sigma\alpha}}\right] + C^t_\alpha(\neg p) \times E\left[\tau^t_{\sigma\alpha}\right]}.$$

Since we, by the requirement of conditionalization, must have that $C^{t+1}_\alpha(p) = C^t_\alpha\left(p | S^t_{\sigma\alpha}(p)\right)$ whenever σ is the only source giving information to α at t, and that information consists in the message p, these formulas completely determine how α should update her credences in such a case. Some of the consequences of this can be summarized qualitatively. We say that σ is *trusted* when $E\left[\tau^t_{\sigma\alpha}\right] > 0.5$, *distrusted* when $E\left[\tau^t_{\sigma\alpha}\right] < 0.5$, and neither trusted nor distrusted when $E\left[\tau^t_{\sigma\alpha}\right] = 0.5$. Furthermore, we say that a message m is *plausible* if $C^t_\alpha(p) > 0.5$ and $m \equiv p$ or $C^t_\alpha(p) < 0.5$ and $m \equiv \neg p$,

[5] We are indebted to an anonymous referee for pressing us on the exact formulation of our version of the principal principle in terms of intervals rather than point values.

Table 2.1 **Effects of Trust on Credence**

Is source trusted?	*Is message plausible?*		
	Yes	*Neither*	*No*
Yes	+	+	
Neither	0	0	0
No	–	–	+

implausible if $C^t_\alpha(p) < 0.5$ and $m \equiv p$ or $C^t_\alpha(p) > 0.5$ and $m \equiv \neg p$, and neither plausible nor implausible otherwise. Table 2.1 (see Vallinder and Olsson 2013 for derivations and Collins et al. 2015 for some empirical support) gives the qualitative rules for how belief is updated. A "+" means that the message reinforces α's current belief (i.e., her credence increases if above 0.5 and decreases if below 0.5), a "–" that the strength of her belief is weakened (i.e. that her credence increases if below 0.5 and decreases if above 0.5), and "0" that her credence is left unchanged.

The calculations become slightly more complex when we take into account the possibility of receiving several messages at the same time. Let Σ^t_α be the set of sources from which α receives information at t, and let $m^t_{\sigma\alpha}$ be the message (i.e., either p or not-p) that σ gives α at t. Conditionalization requires that

$$C^{t+1}_\alpha(p) = C^t_\alpha\left(p \mid \vee_{\sigma\in\Sigma^t_\alpha} S^t_{\sigma\alpha}\left(m^t_{\sigma\alpha}\right)\right).$$

This may be a rather complex expression, but we can simplify greatly it if we take inquirers to treat their sources as independent given the truth or falsity of p. Formally, this means that we adopt the following axiom, which we call *source independence* (*SI*):

$$C^t_\alpha\left(\vee_{\sigma\in\Sigma^t_\alpha} S^t_{\sigma\alpha}\left(m^t_{\sigma\alpha}\right)\mid p\right) = \prod_{\sigma\in\Sigma^t_\alpha} C^t_\alpha(S^t_{\sigma\alpha}\left(m^t_{\sigma\alpha}\right)\mid p).$$

Is this a valid assumption to make? In certain cases, it may very well be, at least as an approximation. We generally tend to exclude conspiracy or other forms of dependence in the absence of concrete evidence to the contrary. It is this default assumption of independence that makes us attach a greater degree of belief to information coming from different sources. Although the assumption relies on a simplification—there may be all kinds of hidden dependencies of which we are unaware—acting on the basis of it may be the best thing to do, all things considered.

We should note that communication dependence does not contradict the fact that the credences of inquirers indeed do become probabilistically dependent as a result of their interactions with one another. This is because *SI* applies to credences rather than the probabilities for having these credences. The result is, however, that as the simulation proceeds, the assumption of source independence implies that the inquirers themselves will become less and less informed about the current structure of the network. As noted in the previous paragraph, this is a simplification. However, it is one that is necessary in order to limit to manageable levels the information inquirers must store.

Given source independence, the properties of individual links and inquirers determine how inquirers should update their beliefs when they receive new information.

An application of Bayes' theorem together with the theorem of total probability gives that

$$C_\alpha^t\left(\vee_{\sigma\in\Sigma_\alpha^t} S_{\sigma\alpha}^t\left(m_{\sigma\alpha}^t\right)|p\right)$$
$$=\frac{C_\alpha^t(p)\times\prod_{\sigma\in\Sigma_\alpha^t} C_\alpha^t(S_{\sigma\alpha}^t\left(m_{\sigma\alpha}^t\right)|p)}{C_\alpha^t(p)\times\prod_{\sigma\in\Sigma_\alpha^t} C_\alpha^t\left(S_{\sigma\alpha}^t\left(m_{\sigma\alpha}^t\right)|p\right)+C_\alpha^t(\neg p)\times\prod_{\sigma\in\Sigma_\alpha^t} C_\alpha^t(S_{\sigma\alpha}^t\left(m_{\sigma\alpha}^t\right)|\neg p)}.$$

Since the values $C_\alpha^t(S_{\sigma\alpha}^t\left(m_{\sigma\alpha}^t\right)|p)$ and $C_\alpha^t(S_{\sigma\alpha}^t\left(m_{\sigma\alpha}^t\right)|\neg p)$ are determined by equation (2.1), this equation lets us infer exactly what degree of belief an inquirer α should give p, given the information she receives at any time.

However, α's credence in p is not the only part of her epistemic state that needs to be updated in light of new evidence. Equally important is for α to keep track of how much to trust her sources. A source that generally gives very implausible reports is unlikely to be veridical, and an inquirer should adjust her trust function in light of this. It turns out that our model already determines how to do this: given that α's trust function for the source σ at t is $\tau_{\sigma\alpha}^t$, and that she receives the result that p or that not-p from σ then, her new trust function $\tau_{\sigma\alpha}^{t+1}$ must be given either by

$$\tau_{\sigma\alpha}^{t+1}(r)=\tau_{\sigma\alpha}^t(r)\times\frac{r\times C_\alpha^t(p)+\bar{r}\times C_\alpha^t(\neg p)}{E\left[\tau_{\sigma\alpha}^t\right]\times C_\alpha^t(p)+E\left[\overline{\tau_{\sigma\alpha}^t}\right]\times C_\alpha^t(\neg p)}$$

or by

$$\tau_{\sigma\alpha}^{t+1}(r)=\tau_{\sigma\alpha}^t(r)\times\frac{r\times C_\alpha^t(\neg p)+\bar{r}\times C_\alpha^t(p)}{E\left[\tau_{\sigma\alpha}^t\right]\times C_\alpha^t(\neg p)+E\left[\overline{\tau_{\sigma\alpha}^t}\right]\times C_\alpha^t(p)}$$

depending on whether the report received from σ claims that p or that not-p.[6]

That both $C_\alpha^t(p)$ and $\tau_{\sigma\alpha}^t$ change as a result of inquiry and communication gives rise to complex interactions. Assume, for example, that α starts out with a credence in p slightly below 0.14, and some reasonable degree of trust in her own inquiry, indicated by a value of expected trust at 0.67. Assume that her inquiry keeps indicating that p actually is the case. How her trust in inquiry changes after t such inputs depends crucially on the trust function τ she starts out with. But even if the *expected* trust is the same, the shape of the trust function influences the long-term behavior. Figure 2.2 illustrates three such scenarios during a time lapse of 50 inputs. All the scenarios start out with trust functions having the form of beta distributions with different values of α and β, although with the same expectation.

Here we have three different ways that the inquirer's credence might evolve given the same initial credence, the same obtained evidence, and the same initial expected trust. Although our inquirer happens to be a perfect inquirer insofar as her inquiry always gives the right result, the fairly low stability of her faith in inquiry in (*a*), together with her prior judgment that p is unlikely, conspire to make her distrust her own inquiry. This, in turn, gives rise to a vicious circle in which she becomes more and more convinced that p is false, and that her inquiry is negatively correlated with the truth.

6 Again, see the appendix for the derivation.

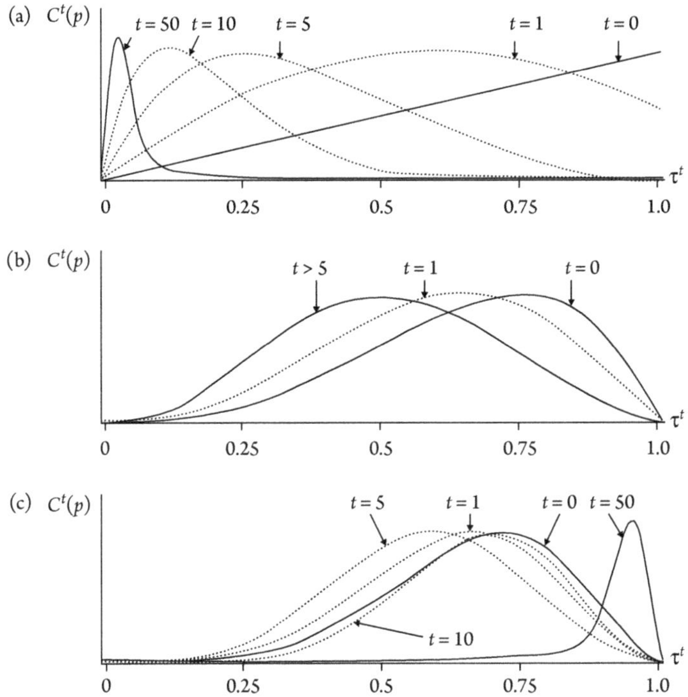

Figure 2.2 Results of iterated trust updates

In (*b*), the inquirer's trust happens to be just enough to counter her prior disbelief in p, but not enough to get her to believe that p, with the result that her credence converges to 0.5. In (*c*), her trust in inquiry is stable enough to overcome her prior disbelief, with the happy consequence that she converges on the truth.

What can we learn from these scenarios? The most important lesson is that trust is a complex issue, impossible to capture in a single number. Although we mentioned 'stability' in the preceding paragraphs, it should be noted that two numbers are still insufficient, so stability should not be seen as a complement to (expected) trust. Although only the expected trust influences an inquirer's posterior credence for a single update, this does not hold for further updates. For arbitrarily iterated updating, all of τ is necessary.

2.5 Setting Up the Experiment

We have given a quick overview of the model we will be using to describe a research network. This model has been implemented in the simulation environment Laputa, which allows us to run simulations of such networks and collect their statistics.

For the present investigation, two of the most important properties are the *density* of the network and the certainty threshold. The density of a graph (and a research network) is the average number of edges (links) per ordered pair of inquirers, excluding pairs with the same first and second element. In a network with n inquirers and m links, the density will therefore be

$$\delta = \frac{m}{n \times (n-1)}.$$

Laputa can be set to generate networks with a specified density but with different characteristics. The most straightforward method of generating a network of density δ is to simply go through each ordered pair of inquirers and add a link between the members of each pair with probability δ. Networks obtained this way are called Erdős-Renyi graphs, and have been studied intensely. The edge distribution of an Erdős-Renyi graph is binomial, i.e., plotting a histogram of the number of edges connecting a node gives rise to a binomial distribution.

However, real networks of scientific collaborations, as well as networks of citations, tend to have a different structure, which shows up in the fact that the tails of their edge distributions conform to a power law instead of a binomial distribution.[7] One reason for this is the "rich get richer" mechanism, which makes nodes with more edges proportionally more likely to attract new edges than nodes with less. In terms of citations, papers that are cited more tend to be read more, and thus again be cited more. Laputa allows taking this effect into account, and we have done so by assigning inquirers with n links going out of them with weight $n+1$, rather than 1, which would have been the case in the Erdős-Renyi model.[8] This approximates the model of Barabási and Albert (1999).

Together with the communication chance of each link, the density δ gives a good indication of how much communication is going on in a research network. Since simulations with fewer links run faster than simulations with more links, but lower communication chance, we have opted to vary δ and keep the communication chance random.

When $\delta = 0$, no communication at all occurs, and when $\delta = 1$, every inquirer can communicate directly with everyone else. However, this does not mean that they actually *do* communicate, and one of the variables that controls whether this happens is the certainty threshold. $\theta_{\sigma\alpha}$ is part of what may be called the network's *norms of assertion*: it poses a requirement on when p can be communicated in terms of the source's epistemic state (Olsson and Vallinder 2013). Investigating such norms of assertion is one of the objectives of the present study. For simplicity, we assume that all links share a common certainty threshold θ.

The other variables of the model are among those we are not currently interested in. Dealing with them properly requires us to randomize them so as to obtain a *randomized experiment*, which will allow us to draw conclusions about the *causal* effect of δ and θ on a network's performance. The specific random distributions that the other variables are picked from should be dependent on the model's intended applications.

We pick the starting degree of belief for each inquirer from a normal distribution with mean 0.5 and standard deviation 0.15. This corresponds to the intended applications being research networks in which the starting degrees of belief are somewhat

[7] See, for example, Newman (2001) and Redner (1998).

[8] This is a novelty in Laputa v1.5 of December 2013. Earlier versions lacked this feature, and allowed the generation of Erdős-Renyi networks only.

clustered around 0.5: like good Bayesians, our researchers do not usually start out with strong beliefs about whether or not p is the case.

Both the inquiry chance and the link communication chance are uniformly distributed in [0, 1]. This means that we take each degree of activity, of inquirers qua researchers of links qua communication channels, to be equally likely.

We will study two possible distributions of inquiry accuracy: the first will be a normal distribution with mean 0.55 and standard deviation 0.15. This gives us inquirers that are, on average, slightly better than chance. The second distribution is also a normal distribution with the same variance, but with mean 0.75, which is intended to describe research networks in which inquirers tend to be highly competent.

Both inquiry trust and the trust functions of each link are set to normal distributions with means randomly picked between 0.5 and 1.0, and standard deviations randomly picked between 1 and 0.15. Hence inquirers will generally start out trusting their inquiry to some degree, and also trusting other inquirers to some degree. This represents a basic form of the principle of charity, which is a reasonable assumption to make when modeling real scientific communities.

To begin with, we are mainly interested in the end result of inquiry, so we let each randomized research group run for 25 steps, which is sufficient for inquirers' opinions to stabilize. Each research group is taken to have 25 members, or, alternatively, we can see it as having from 1 to 25 members, with the number of members weighted by the number of possible graphs of that order, which gives us a uniform distribution over all graphs of order ≤ 25. Since the number of directed graphs of order $n+1$ is 2^{2n} times that of order n, the difference between considering only groups of 25 inquirers and groups of 1 to 25 inquirers is negligible.[9]

This concludes our specification of the parameters of the model. What we still have to describe is the way we wish to measure a network's performance. Goldman (1999) introduces the notion of *veritistic value*, or *V-value*, of a social practice. For the single-proposition case, which is what we are working with, this is calculated as the network's average increase in the credence in p, where p is taken to be true. The proposition p will be assumed true in all the following simulations.

In our case, a *social practice* can be identified with a way of constraining possible evolutions of a research group as it engages in inquiry (Olsson 2011). But since such evolutions are determined, at least probabilistically, by their initial states, we can take a social practice to be so determined as well. Thus each setup of the model we have (or in Laputa terms, each batch simulation) gives a social practice in this extended sense. We denote the V-value of a network of inquirers N at a given time $V^t(N)$.

V-value gives a straight mean of the influence of a practice, but, just as the GDP per capita of a country says nothing about its distribution of wealth, V-value says nothing

[9] Sampling uniformly from *isomorphism classes* of graphs is a significantly harder problem. However, we still have that there are about 6.7×10^6 more isomorphism classes of graphs with 25 vertices than with 24, and 2.3×10^{11} more than with 23 (see http://oeis.org/A000088), so just picking from graphs with 25 vertices still gives a very good approximation. See Masterton (2014) for an extensive investigation into the graph isomorphism issue as it applies to Laputa and related frameworks.

about the distribution of knowledge. It is therefore useful to also plot the *polarization* of belief that follows from a given practice, by which we mean the root mean square of the deviation of individual credence from the mean, i.e.,

$$Pol^t(N) = \sqrt{\frac{1}{|N|} \sum_{\alpha \in N} (C^t_\alpha(p) - \mu^t)^2},$$

where $\mu^t = \frac{1}{|N|} \sum_{C^t_\alpha(p)} C^t_\alpha(p)$. A network in which every inquirer has the same credence in p will thus have minimum polarization 0, while one in which half the inquirers are certain that p and half are certain that not–p will have maximum polarization 0.5.

A different limitation of V-value is its relative irrelevance for an important type of group decision-making: from the V-value of a practice, little can be inferred about the result of a majority vote among the members of a group that have just implemented that practice. We will therefore also measure the *average belief of the majority*, which we define as the average over all simulation runs of the value

$$BM^t(N) = \begin{cases} 1 \text{ if } & \left|\{\alpha \in N | V^t(\alpha) > 0.5\}\right| > \left|\{\alpha \in N | V^t(\alpha) < 0.5\}\right| \\ 0 \text{ if } & \left|\{\alpha \in N | V^t(\alpha) > 0.5\}\right| < \left|\{\alpha \in N | V^t(\alpha) < 0.5\}\right| \end{cases}.$$

The average belief of the majority (or ABM) that we are interested in is, like V-value, that which is taken by N in its *final* states, i.e., after the simulation has run its course of 25 iterations.

2.6 The Perils of Unrestricted Communication

Running the simulation for 25,000 iterations in order to achieve more numerically stable results, using the parameters described in the previous section, produced the results shown in figure 2.3.

We can see that, in all cases, *more communication is bad*, both in terms of V-value and ABM. The threshold does not affect either V-value or ABM, and while the inquirers' accuracy does, it does not make any amount of communication preferable to none at all. Whether inquirer as slightly or highly accurate, no communication at all gives a V-value approximately 5 times higher than full communication.

Note that this does not occur because inquirers in highly dense networks engage in communication *instead of* inquiry: in Laputa, whether a link is used during a time step is independent of whether the source also engages in inquiry during that time. As we mentioned in the introduction, we thus have a puzzle: scientific communication is usually taken for granted to be a good thing, but we have just found an indication that it is not. Why? What is so bad about communication that it makes inquirers less likely to converge on full credence in the truth?

Let us first ask why one might think that groups should perform better than single inquirers in the first place. One reason is the famous Condorcet jury theorem, which states that the majority vote is more reliable than the vote of any given voter under favorable circumstances. This theorem applies to majority belief as a kind of collective "vote," but it does not have any obvious implications for V-value. However, even with regard to majority belief, the theorem has an important presupposition: the voters

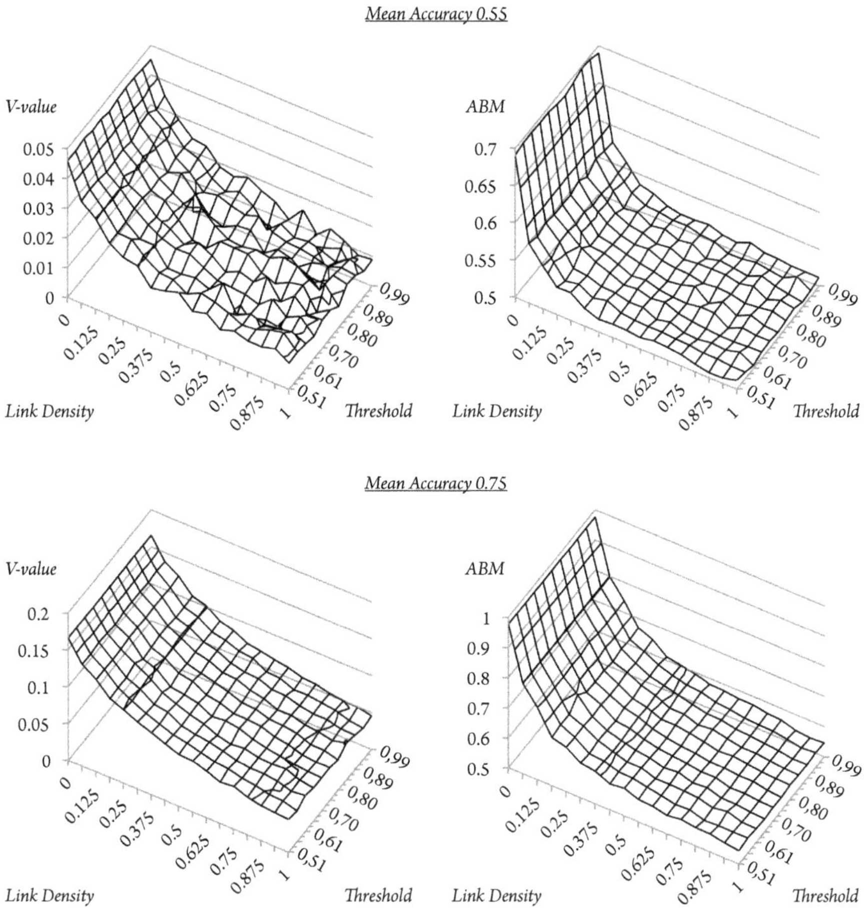

Figure 2.3 Results of unrestricted communication

in question have to be *independent*. If they are able to directly influence one another, the theorem no longer applies. Now scientific communication is a prime example of such influence, and we propose that the downside to communication that makes the results of figure 2.3 possible is that communication can, and typically does, *reduce independence*. This is particularly problematic when the initial credences are distributed around 0.5 as we have, realistically, assumed them to be. The denser the network, the more sensitive it will be to misplaced initial credences, i.e., initial credences that lie slightly below 0.5. If the majority starts off on the wrong track, chances are that it will drag down the whole network. For inquirers will then receive a strong social signal that not-*p* is true when in fact we have assumed *p* to be true and hence not-*p* to be false. This in turn may lead inquirers to downgrade their trust in their own inquiry to the point where they consider it unreliable, preventing self-correction through inquiry from taking place. We would expect the probability of this unfortunate development to increase with the network density.

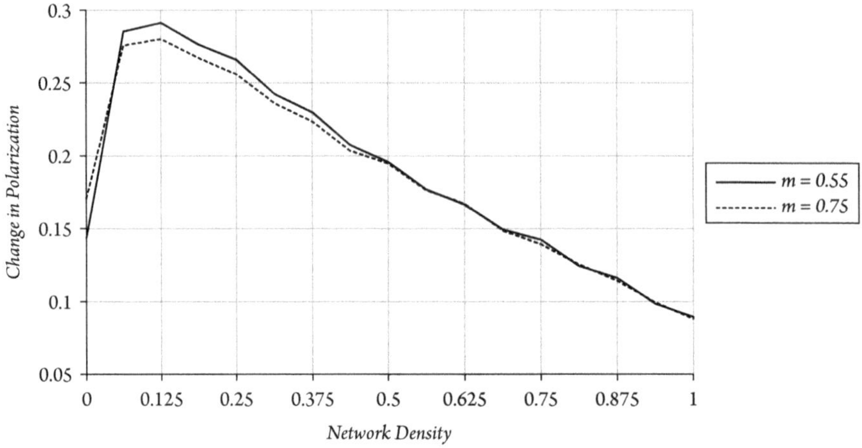

Figure 2.4 Change in polarization as a result of the practices in figure 2.2

There is an upside of sorts to this, though. The diagram in figure 2.4 shows the degree of polarization as a result of the previous practices. Since the numbers, like those of the figure 2.2, are independent of the threshold, we have plotted the change in polarizations of both values of mean accuracy (*m*) as lines, rather than as surfaces.

As seen in the diagram, a degree of communication over zero but below about 0.65 increases the polarization of belief in a research network. On the other hand, high network density decreases it, so while a large amount of communication decreases the V-value, it has an egalitarian aspect: it makes differences of opinion on *p* less pronounced. Whether this should count as an *epistemic* advantage may be left unsaid; presumably it can be a practical one.

It is worth pointing out that since the change in polarization is above zero for *all* density values, the process of inquiry generally has the effect of diminishing differences of opinion. The reason for this phenomenon is our handling of trust, which not only models how trust affects belief, but also how belief affects trust. In the typical case, inquirers will start out with beliefs around 0.5, but by chance some get results that not-*p* from inquiry, either because they have a low accuracy or because of bad luck. Since they typically trust their own inquiry somewhat, this will make them start to believe that not-*p* is the case. If they happen to receive strong enough confirmation that not-*p*, messages that *p* from others will be taken as indications of untrustworthiness, and will thus only further reinforce the conviction that not-*p*.

These conclusions may seem somewhat pessimistic, but they follow from a Bayesian approach together with the axioms we have imposed, and the modeling of trust used. Unlike traditional Bayesians, who rely on the Bernstein–von Mises theorem or some other convergence result in order to show that convergence has to occur, we recognize that prior probabilities influence not only our beliefs about the facts, but also our interpretation of the evidence, i.e., the perceived reliability of our inquiries. This means that we can, and frequently will, get divergences of opinion even in the long run, and that this is not due to irrationality but rather exactly what rationality prescribes.

As we saw, increasing the certainty threshold does not increase either V-value or ABM. But are there other restrictions to communication that could make it more effective? At present, whether a link can be used depends only on chance and the threshold, but this means that inquirers can *spam* the network by having many outgoing links with high communication chance. Such inquirers will transmit their opinion repeatedly. If they start out with initial credences that tend toward not-p rather than p, which, as we saw, is a serious possibility given our assumptions, there is a fair chance that the false information is repeatedly transmitted in the network. And, as we also noted, the impact of massively false social information on a given inquirer may very well be that she starts distrusting her own inquiry, i.e., starts to regard it as unreliable, in which case the probability of self-correction through inquiry is severely reduced.

Stopping this form of recycled information would require asking inquirers not to transmit their opinion across a link unless they have actually received new supporting information since that link was last used. We will consider two conditions along this line:

> *New Evidence* (*NE*): To use link $\beta\alpha$ to transmit p (not-p) one of the following must be satisfied:
>
> (a) $\beta\alpha$ has not been used before, so that the information transmitted is based on β's prior credence in p (not-p).
> (b) Since the last time $\beta\alpha$ was used β has received a message that p (not-p) from some inquirer other than α, which has resulted in her increasing her credence in p (not-p).
> (c) Since the last time $\beta\alpha$ was used β has received a result that p (not-p) from her own inquiry, which has resulted in her increasing her credence in p (not-p).
>
> *New Inquiry* (*NI*): To use link $\beta\alpha$ to transmit p (not-p), condition (c) of *NE* must be satisfied.

Imposing *NE* or *NI* means that an inquirer cannot retransmit information in the network unless she has obtained evidence in support of that information since the last time she transmitted it over the same link. Intuitively, the effect should be to prevent insufficiently supported information from spreading in a massively connected network. This is especially clear in the case of *NI*. To be in a position to retransmit not-p the inquirer would, if *NI* is imposed, have to obtain not-p from inquiry, which, given the assumed reliability of inquiry, is relatively unlikely.

From another perspective, imposing *NE* or *NI* can be seen as a partial fix of one limitation of the model: that the only information transmittable by inquirers in the network is p or not-p, and no explicit discussion about evidence occurs. With *NE* or *NI* in place, a message that p can be interpreted as 'there is new evidence that p' or 'I made a new observation of p'. This gives us a simple way to accommodate evidence in the model.[10]

[10] We should, however, recognize that *NE* and *NI* are only partial fixes, or in programmers' jargon, "hacks." They do not amount to making inquirers actually keep track of particular pieces of evidence. For example, *NE* forbids two inquirers from strengthening each other's beliefs by just

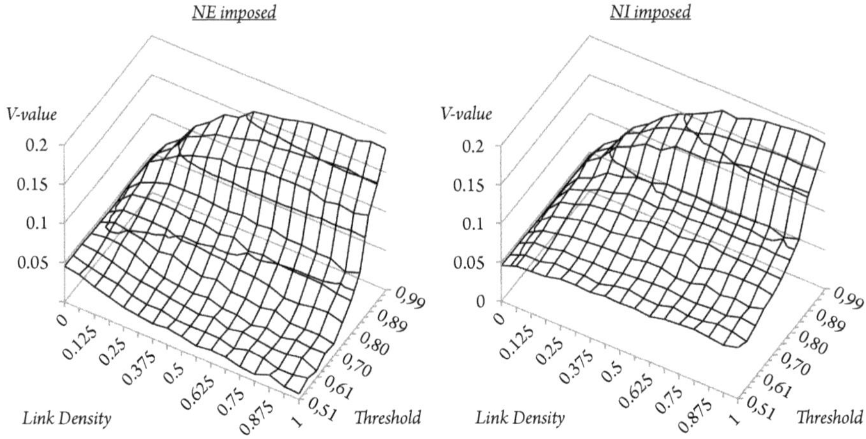

Figure 2.5 Impact on V-value of imposing *NE* and *NI*

2.7 Effects of the Quality Constraints

Imposing *NE* or *NI* gives the *V*-values shown in figure 2.5, for inquirers with mean accuracy 0.55.[11]

With the restrictions in place, communication can actually be beneficial in terms of *V*-value (figure 2.5). With a high certainty threshold, the best network density values lie around 0.5 for *NE*, and for *NI* we observe that more communication *never* decreases *V*-value. Furthermore, the V-values for *NI* are higher than those for *NE*, so among the two, the stricter condition *NI* is preferable.

NE and *NI* also have interesting consequences for belief polarization. While the threshold had no effect for unrestricted communication, its value is important for the restricted case, as seen in figure 2.6.

We see that, just as for unrestricted communication, a small amount of communication makes the polarization rise, and increasing it further makes it fall again. But the amount it falls depends on the threshold, and using *NE* and setting the threshold at around 0.8 and the density to 1, a minimum of belief polarization is attained.

Taken together, the diagrams for V-value and polarization (figures 2.5 and 2.6, respectively) imply a conflict concerning the proper assertion threshold. From the perspective of minimizing polarization, the threshold should be at around 0.80, but V-value is maximized when the threshold is 0.99. There is also a conflict as to what restriction to use: for polarization, *NE* is by far the best, while for V-value, *NI* is somewhat better.

The threshold is also important for the effect of *NE* and *NI* on the average belief of the majority (ABM), as seen in figure 2.7.

talking among themselves, but does not do this for three or more inquirers connected in a circle. The problem of extending the model to take evidence into account in a more robust way that excludes such cases as well remains to be solved.

[11] Very similar results were obtained for the case of mean accuracy 0.75, although, as before, with higher V-values and ABM. To conserve space, we have not reproduced them here.

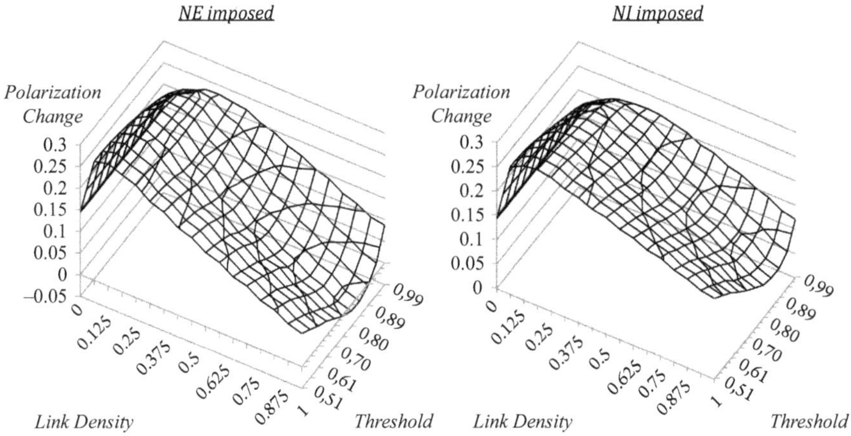

Figure 2.6 Impact on belief polarization of imposing *NE* and *NI*

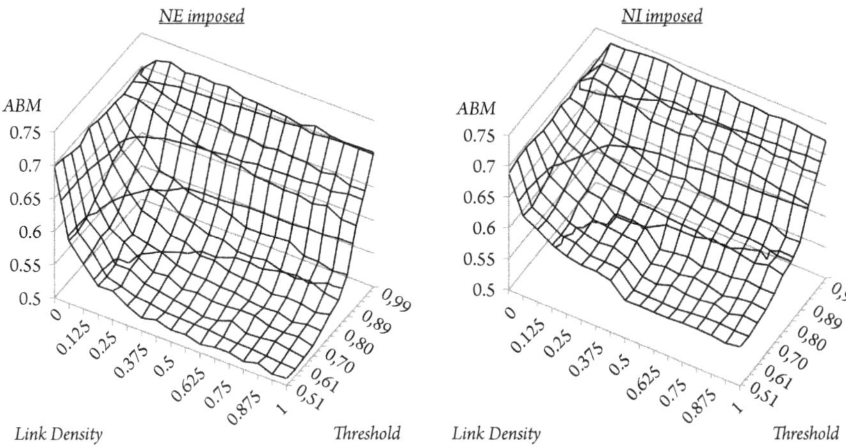

Figure 2.7 Impact on average belief of the majority of imposing *NE* and *NI*

In terms of ABM, we find that it is still generally the case that the best policy is for researchers not to communicate at all. The only exception is for very high values of the threshold, where a small increase is somewhat beneficial. This is noteworthy because the belief of the majority is arguably more important than V-value in democratically organized groups, where many important decisions are decided through a majority vote. In particular, the results we have obtained seem to count against important assumptions of deliberative democracy, in which group deliberation prior to voting is taken to be beneficial for the result.[12] Our study indicates that, at least epistemically, it would be better if no communication occurred prior to voting.

Olsson and Vallinder (2013) make a related observation about V-value, also using Laputa. As we noted, in the absence of restrictions, communication generally decreases

12 Cf. Goodin (2003).

the V-value in the long run. But in the short run, Olsson and Vallinder show, communication *can* increase the V-value. While imposing *NE* or *NI* secures a V-value increase even in the long run, Olsson and Vallinder's general observation carries over. The diagrams in figure 2.8 show the V-value of network different density values, plotted against the number of simulation steps. The threshold has been set at 0.95.

Both with *NE* and *NI*, higher density of the network is clearly beneficial in the V-value sense. In fact, more communication makes the society converge to its final V-value quicker, *as well as* making that V-value higher. *NI* is, on the whole, somewhat better than *NE*, but the differences are fairly small.

Given that shortening the deliberation time makes higher network density lead to higher V-value, it is natural to ask if the same is the case for ABM. Unfortunately, as we can see from the diagram in figure 2.9, the effects of lowering the number of simulation steps are rather unimpressive.

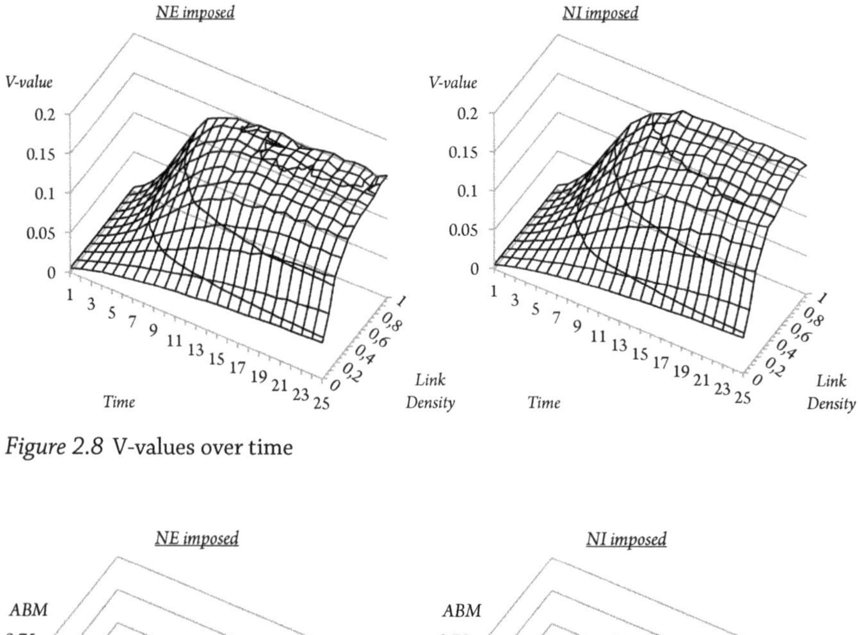

Figure 2.8 V-values over time

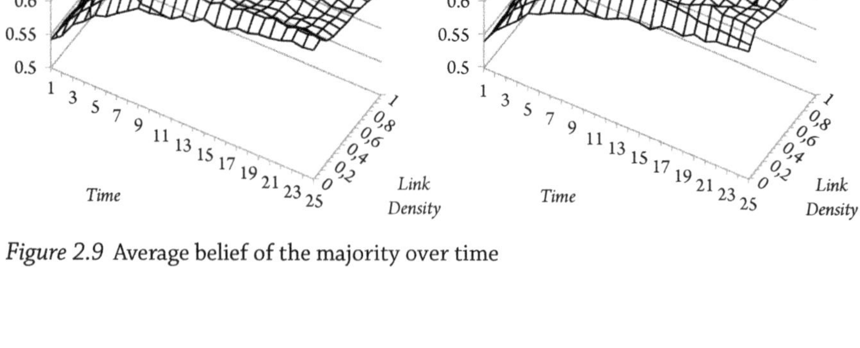

Figure 2.9 Average belief of the majority over time

As in the long run, there is a small beneficial effect to having some communication rather than none at all, but this effect is actually smaller for short simulations rather than larger. Thus we have as yet no case where a high amount of communication would increase the accuracy of the majority opinion. For a proponent of deliberative democracy, it might be natural to blame Laputa's simplified model of communication for this effect. Real research networks communicate not only their beliefs, but also the *reasons* for those beliefs. But with the restrictions in place, it is hard to see why this feature should affect the results substantially. Especially if *NI* is adopted, any communication requires new (supporting) results from inquiry, i.e., new reasons (cf. Olsson 2013).

2.8 Discussion and Conclusion

As noted in Vallinder and Olsson (2013), the Laputa model can, being Bayesian, be given a normative interpretation. Olsson (2013) is an extended discussion of the normative and descriptive interpretations of Laputa, connecting it to the empirically robust Persuasive Argument Theory (PAT) in social psychology, and suggests that Laputa is largely subsumable under the latter. In addition, Olsson shows that inquirers in Laputa survive what he calls the "polarization test": "if initially disposed to judge along the same lines, inquirers in Laputa will adopt a more extreme position in the same direction as the effect of group deliberation, just like members of real argumentative bodies" (p. 113).

The influential simulation model developed in Hegselmann and Krause (2006), the HK-model for short, was one source of inspiration behind Laputa. In Vallinder and Olsson (2013), it was suggested that Laputa and the HK-model are competitors among simulation models that are truth sensitive. The HK-model has been developed in various directions by other researchers and put to interesting uses in various philosophical applications (e.g., Douven and Kelp 2011). However, the HK-model does not as naturally lend itself to the study of the influence of network density on group competence per se, since the network is not explicitly represented in the model.[13]

A stable result of our investigation is that it is generally beneficial, in terms of V-value, to require inquirers to show some discretion concerning what information to post in the network. Casually asserted or recycled information tends to pull the entire network away from truth, and evidential norms like the ones we have suggested make this less likely to happen. What comes out of our simulations, regarding V-value, is that adhering to such norms becomes increasingly important as scientists become ever more connected. Network density can be positively correlated with group performance if inquirers are exercising proper restraint in their communicational behavior.

The present chapter bears on a matter addressed in Olsson (2013, p. 124 n. 3): "We have been experimenting with a version of the [Laputa] program in which

[13] As Rainer Hegselmann has pointed out to us (personal communication) there are various ways of (re)interpreting the HK-model in terms of a network structure. Unfortunately, a fuller investigation is outside the scope of the present article. A comparison between the two frameworks from another perspective can be found in Vallinder and Olsson (2013), p. 1441.

communication is possible only if new inquiry has taken place. Preliminary simulations suggest that this modification does not have any significant statistical effect on simulation outcome." The more extensive and systematic simulations carried out in the present article disprove this suggestion. By prohibiting arbitrary repetition of information in the network, the present version of Laputa makes it possible to enforce an "argumentation interpretation" of the model, as defended in Olsson's 2013 article and further detailed in Masterton and Olsson (2013).

To be more precise, in terms of V-value as well as polarization, it is, as long as our restrictions are in effect, best for everyone to be able to communicate with everyone else. But whereas the best effect regarding V-value is obtained for a threshold of 0.99, the optimum level of polarization is obtained for a threshold of 0.80. So how we should set the threshold depends on what we value the most: having accurate credence or avoiding polarization. In this case, an epistemic value (having accurate credences) conflicts with a value of a practical kind (avoiding polarization). We also get different results depending on what kind of *epistemic* value we take to be most central. In terms of average belief of the majority, the best course of action is not to communicate at all, or to do so rarely. This remains true even if our restrictions on communication are imposed. So, again, there is a conflict in what practice should be adopted, depending on what we value the most, but this time the conflicting values are both of an epistemic nature.[14]

Finally, we would like to mention a particularly suggestive aspect of our study that may be of potential practical significance, namely, the fact that it seems to create doubt on the recommendation to "publish early, publish often" frequently given to young scholars as key to a successful academic career. To see how our study is relevant, recall that our results indicate that unless inquires agree on certain restrictions on their communications, including scientific publications, an increasingly networked scientific community runs an ever greater risk of betting on the wrong theory, as it were. We went on to show how this effect can be counteracted by scientists agreeing (i) to communicate only information judged to be practically certain and (ii) not to recycle information in the absence of new supporting evidence.

It seems to us that the said advice is difficult to reconcile with these two restrictions. In order to publish early and often one would normally have to publish things that one has doubts about. A certain amount of recycling of old material also seems necessary in practice if one wants to publish often, as readers can surely testify based on their own experience. Our study indicates that, as the scientific community becomes increasingly connected, scholars should, in the interest of science as an institution, be increasingly self-critical when deciding whether to publish or not, even if this means publishing late and rarely. Alternatively or complementarily, journals and other scientific gatekeepers should be increasingly unwilling to publish doubtful or recycled articles. Unfortunately, precisely the opposite is arguably happening at the

[14] We are greatly indebted to two anonymous referees for several suggestions for improvements. We have also benefited from comments by Rainer Hegselmann on interpretational issues and, in particular, on the relation between his model and ours. Our work on this article was supported by a grant from the Swedish Research Council (Collective Competence in Deliberative Groups: On the Epistemological Foundations of Democracy).

time of writing (July 2016). While the scientific community is becoming increasingly intertwined through electronic communication, there is also a trend of publishing in more or less obscure open access journals whose function as scientific gatekeepers can be questioned. Our study gives reasons to believe that the combined effect of increased networking and less than responsible open access publishing may be more problematic than has previously been assumed.

Appendix

Derivation of the Credence and Trust Update Functions

We will need the following assumptions for the derivation of the credence function:

Principal Principle (PP):

$$a < C^t_\alpha \left(S^t_{\sigma\alpha}(p) \mid S^t_{\sigma\alpha}, p, a < r_{\sigma\alpha} < b\right) < b$$
$$a < C^t_\alpha \left(S^t_{\sigma\alpha}(\neg p) \mid S^t_{\sigma\alpha}, \neg p, a < r_{\sigma\alpha} < b\right) < b$$

Communication Independence (CI):

$$C^t_\alpha \left(S^t_{\sigma\alpha}, p, a < r_{\sigma\alpha} < b\right) = C^t_\alpha \left(S^t_{\sigma\alpha}\right) \times C^t_\alpha (p) \times C^t_\alpha (a < r_{\sigma\alpha} < b)$$
$$C^t_\alpha \left(S^t_{\sigma\alpha}, \neg p, a < r_{\sigma\alpha} < b\right) = C^t_\alpha \left(S^t_{\sigma\alpha}\right) \times C^t_\alpha (\neg p) \times C^t_\alpha (a < r_{\sigma\alpha} < b)$$

Source Independence (SI):

$$C^t_\alpha \left(\vee_{\sigma \in \Sigma^t_\alpha} S^t_{\sigma\alpha}\left(m^t_{\sigma\alpha}\right) | p\right) = \prod_{\sigma \in \Sigma^t_\alpha} C^t_\alpha (S^t_{\sigma\alpha}\left(m^t_{\sigma\alpha}\right) | p).$$

$$C^t_\alpha \left(\vee_{\sigma \in \Sigma^t_\alpha} S^t_{\sigma\alpha}\left(m^t_{\sigma\alpha}\right) | \neg p\right) = \prod_{\sigma \in \Sigma^t_\alpha} C^t_\alpha (S^t_{\sigma\alpha}\left(m^t_{\sigma\alpha}\right) | \neg p),$$

where $0 \leq a < b \leq 1$, Σ^t_α is the set of sources that give information to α at t, and $m^t_{\sigma\alpha}$ is the content of the source σ's message.

Since the trust function, which plays a crucial part in the model, is continuous, the derivation will sometimes need to take a detour through conditional probability densities rather than the conditional probabilities themselves. We will briefly sketch how this can be done here.

We have so far not been specific about the σ-algebra Z that C^t_α is defined on. Assume that it is product of several such algebras, the first of which is discrete and generated by atomic events such as p, $\neg p$, $S_{\beta\alpha}(p)$ etc., and the others, which are continuous, are generated by events of the form $a \leq r_{\sigma\alpha} \leq b$. Call the first algebra X and the others $Y_{\sigma_0}, \ldots, Y_{\sigma_n}$. It is clear that, as long as time and the number of inquirers are both finite, X will have only finitely many elements. On the other hand, $Y_{\sigma_0}, \ldots, Y_{\sigma_n}$ are certainly infinite. As mentioned, we assume that $Z = X \times Y_{\sigma_0} \times \cdots \times Y_{\sigma_n}$. Given any source σ_k and time t, we can therefore interpret the part of C^t_α defined on the subalgebra

$X \times Y_{\sigma_k}$ of Z as arising from a joint density function $\kappa^{\tau}_{\sigma\alpha}(\varphi;x)$ defined through the equation

$$C^t_\alpha(\varphi, a < r_{\sigma\alpha} < b) = \int_a^b \kappa^t_{\sigma\alpha}(\varphi;x)dx.$$

Since we have used the comma to represent conjunction earlier in the chapter, we use a semicolon here to separate the two variables: the first propositional, and the second real-valued. Like τ, this distribution's existence and essential uniqueness are guaranteed by the Radon-Nikodym theorem, and in fact $\tau^t_{\sigma\alpha}$ is the marginal distribution of $\kappa^t_{\sigma\alpha}$ with respect to the reliability variable $r_{\sigma\alpha}$ in question. Since the conditional distribution of a random variable is the joint distribution divided by the marginal distribution of that variable, this means that we have that

$$\kappa^t_{\sigma\alpha}(\varphi \mid x) = \frac{\kappa^t_{\sigma\alpha}(\varphi;x)}{\tau^t_{\sigma\alpha}(x)},$$

which is what will be used to make sense of what it means to conditionalize on $r_{\sigma\alpha}$ having a certain value rather than merely being inside an interval. Setting $r_{\sigma\alpha} = x$, $a = x - \epsilon$ and $b = x + \epsilon$ in *PP* and *CI* and letting $\epsilon \to 0$, we get the versions

$$\begin{aligned} \textit{Principal Principle}_0(PP_{\text{lim}}) : \kappa^t_{\sigma\alpha}\left(S^t_{\sigma\alpha}(p) \mid S^t_{\sigma\alpha}, p; x\right) \\ = \kappa^t_{\sigma\alpha}\left(S^t_{\sigma\alpha}(\neg p) \mid S^t_{\sigma\alpha}, \neg p; x\right) = x \\ \textit{Communication Independence}_0(CI_{\text{lim}}) : \kappa^t_{\sigma\alpha}(S^t_{\sigma\alpha}, p; x) = C^t_\alpha\left(S^t_{\sigma\alpha}\right) C^t_\alpha(p) \times \tau^t_{\sigma\alpha}(x) \\ \kappa^t_{\sigma\alpha}(S^t_{\sigma\alpha}, \neg p; x) = C^t_\alpha\left(S^t_{\sigma\alpha}\right) C^t_\alpha(\neg p) \times \tau^t_{\sigma\alpha}(x) \end{aligned}$$

We can now proceed with the actual derivation. By conditionalization, we must have that $C^{t+1}_\alpha(p)$ is equal to $C^t_\alpha\left(\vee_{\sigma \in \Sigma^t_\alpha} S^t_{\sigma\alpha}\left(m^t_{\sigma\alpha}\right) | p\right)$. Applying Bayes' theorem and then *SI* to this expression gives

$$\begin{aligned} & C^t_\alpha\left(p \mid \vee_{\sigma \in \Sigma^t_\alpha} S^t_{\sigma\alpha}\left(m^t_{\sigma\alpha}\right)\right) \\ & = \frac{C^t_\alpha(p) \times C^t_\alpha\left(\vee_{\sigma \in \Sigma^t_\alpha} S^t_{\sigma\alpha}\left(m^t_{\sigma\alpha}\right) | p\right)}{C^t_\alpha(p) \times C^t_\alpha\left(\vee_{\sigma \in \Sigma^t_\alpha} S^t_{\sigma\alpha}\left(m^t_{\sigma\alpha}\right) | p\right) + C^t_\alpha(\neg p) \times C^t_\alpha\left(\vee_{\sigma \in \Sigma^t_\alpha} S^t_{\sigma\alpha}\left(m^t_{\sigma\alpha}\right) | \neg p\right)} \\ & = \frac{C^t_\alpha(p) \times \prod_{\sigma \in \Sigma^t_\alpha} C^t_\alpha\left(S^t_{\sigma\alpha}\left(m^t_{\sigma\alpha}\right) | p\right)}{C^t_\alpha(p) \times \prod_{\sigma \in \Sigma^t_\alpha} C^t_\alpha\left(S^t_{\sigma\alpha}\left(m^t_{\sigma\alpha}\right) | p\right) + C^t_\alpha(\neg p) \times \prod_{\sigma \in \Sigma^t_\alpha} C^t_\alpha\left(S^t_{\sigma\alpha}\left(m^t_{\sigma\alpha}\right) | \neg p\right)}, \end{aligned}$$

which gives us the posterior credence in terms of the values $C^t_\alpha\left(S^t_{\sigma\alpha}(p) | p\right)$ and $C^t_\alpha\left(S^t_{\sigma\alpha}(\neg p) | \neg p\right)$. Our next task is thus to derive these expressions. Since $S^t_{\sigma\alpha}(p)$ is equivalent to $S^t_{\sigma\alpha}(p) \wedge S^t_{\sigma\alpha}$, it follows that $C^t_\alpha\left(S^t_{\sigma\alpha}(p) | p\right) = C^t_\alpha\left(S^t_{\sigma\alpha}(p), S^t_{\sigma\alpha} | p\right)$. Applying first the definition of conditional probability and then the continuous law of total probability, the definition of conditional probability again, and finally CI_{lim}, we get

$$
\begin{aligned}
C_\alpha^t\left(S_{\sigma\alpha}^t(p) \mid p\right) &= \frac{1}{C_\alpha^t(p)} \times C_\alpha^t\left(S_{\sigma\alpha}^t(p),\, S_{\sigma\alpha}^t, p\right) \\
&= \frac{1}{C_\alpha^t(p)} \times \int_0^1 \kappa_{\sigma\alpha}^t\left(S_{\sigma\alpha}^t(p),\, S_{\sigma\alpha}^t, p;\, x\right) dx \\
&= \frac{1}{C_\alpha^t(p)} \times \int_0^1 \kappa_{\sigma\alpha}^t\left(S_{\sigma\alpha}^t(p) \mid S_{\sigma\alpha}^t, p;\, x\right) \times \kappa_{\sigma\alpha}^t\left(S_{\sigma\alpha}^t, p;\, x\right) dx \\
&= \frac{1}{C_\alpha^t(p)} \times \int_0^1 \kappa_{\sigma\alpha}^t\left(S_{\sigma\alpha}^t(p) \mid S_{\sigma\alpha}^t, p, x\right) \times C_\alpha^t\left(S_{\sigma\alpha}^t\right) \times C_\alpha^t(p) \times \tau_{\sigma\alpha}^t(x)\, dx \\
&= C_\alpha^t\left(S_{\sigma\alpha}^t\right) \times \int_0^1 \kappa_{\sigma\alpha}^t\left(S_{\sigma\alpha}^t(p) \mid S_{\sigma\alpha}^t, p; x\right) \times \tau_{\sigma\alpha}^t(r)\, dx.
\end{aligned}
$$

But PP_{lim} ensures that $\kappa_{\sigma\alpha}^t\left(S_{\sigma\alpha}^t(p) | S_{\sigma\alpha}^t, p; x\right) = x$, so we get

$$
C_\alpha^t\left(S_{\sigma\alpha}^t(p) \mid p\right) = C_\alpha^t\left(S_{\sigma\alpha}^t\right) \times \int_0^1 x \times \tau_{\sigma\alpha}^t(x)\, dx = C_\alpha^t\left(S_{\sigma\alpha}^t\right) \times E\left[\tau_{\sigma\alpha}^t\right].
$$

Parallel derivations give that

$$
\begin{aligned}
C_\alpha^t\left(S_{\sigma\alpha}^t(\neg p) \mid p\right) &= C_\alpha^t\left(S_{\sigma\alpha}^t\right) \times E\left[\overline{\tau_{\sigma\alpha}^t}\right] \\
C_\alpha^t\left(S_{\sigma\alpha}^t(p) \mid \neg p\right) &= C_\alpha^t\left(S_{\sigma\alpha}^t\right) \times E\left[\tau_{\sigma\alpha}^t\right] \\
C_\alpha^t\left(S_{\sigma\alpha}^t(\neg p) \mid \neg p\right) &= C_\alpha^t\left(S_{\sigma\alpha}^t\right) \times E\left[\overline{\tau_{\sigma\alpha}^t}\right]
\end{aligned}
$$

Now let $\Sigma_\alpha^t(p) \subseteq \Sigma_\alpha^t$ be the set of sources that give α the message p at t, and let $\Sigma_\alpha^t(\neg p) = \Sigma_\alpha^t \setminus \Sigma_\alpha^t(p)$. Plugging the above expressions into our earlier result gives the sought-for expression

$$
\begin{aligned}
C_\alpha^{t+1}(p) &= \frac{C_\alpha^t(p) \times \prod_{\sigma \in \Sigma_\alpha^t(p)} C_\alpha^t\left(S_{\sigma\alpha}^t\left(m_{\sigma\alpha}^t\right) \mid p\right)}{C_\alpha^t(p) \times \prod_{\sigma \in \Sigma_\alpha^t(p)} C_\alpha^t\left(S_{\sigma\alpha}^t\right) + C_\alpha^t(\neg p) \times \prod_{\sigma \in \Sigma_\alpha^t(\neg p)} C_\alpha^t\left(S_{\sigma\alpha}^t\left(m_{\sigma\alpha}^t\right) | \neg p\right)} \\
&= \frac{\gamma}{\gamma + \delta}
\end{aligned}
$$

where

$$
\gamma = C_\alpha^t(p) \times \prod_{\sigma \in \Sigma_\alpha^t(p)} \left(C_\alpha^t\left(S_{\sigma\alpha}^t\right) \times E[\tau_{\sigma\alpha}^t]\right) \times \prod_{\sigma \in \Sigma_\alpha^t(\neg p)} \left(C_\alpha^t\left(S_{\sigma\alpha}^t\right) \times E[\overline{\tau_{\sigma\alpha}^t}]\right)
$$

$$
\delta = C_\alpha^t(\neg p) \times \prod_{\sigma \in \Sigma_\alpha^t(p)} \left(C_\alpha^t\left(S_{\sigma\alpha}^t\right) \times E[\overline{\tau_{\sigma\alpha}^t}]\right) \times \prod_{\sigma \in \Sigma_\alpha^t(\neg p)} \left(C_\alpha^t\left(S_{\sigma\alpha}^t\right) \times E[\tau_{\sigma\alpha}^t]\right)
$$

For the derivation of the trust update expression we assume *PP* and *CI*, but not *SI*. The function we wish to derive is

$$
\tau_{\sigma\alpha}^{t+1}(x) = \kappa_{\sigma\alpha}^t\left(x | S_{\sigma\alpha}^t\left(m_{\sigma\alpha}^t\right)\right)
$$

for a source σ of α, and a message $m^t_{\sigma\alpha}$ from that source. Assume that $m^t_{\sigma\alpha} \equiv p$ (the case $m^t_{\sigma\alpha} \equiv \neg p$ is completely symmetrical). Applying the definition of conditional probability, the equivalence $S^t_{\sigma\alpha} \wedge S^t_{\sigma\alpha}(p) \equiv S^t_{\sigma\alpha}(p)$, and the (discrete) law of total probability, we get

$$\kappa^t_{\sigma\alpha}(x|\ S^t_{\sigma\alpha}(p))$$

$$= \frac{\kappa^t_{\sigma\alpha}\left(S^t_{\sigma\alpha}(p);x\right)}{C^t_\alpha\left(S^t_{\sigma\alpha}(p)\right)} = \frac{\kappa^t_{\sigma\alpha}\left(S^t_{\sigma\alpha}(p),\ S^t_{\sigma\alpha};x\right)}{C^t_\alpha\left(S^t_{\sigma\alpha}(p)\right)}$$

$$= \frac{\kappa^t_{\sigma\alpha}\left(S^t_{\sigma\alpha}(p),\ S^t_{\sigma\alpha},p;x\right) + \kappa^t_{\sigma\alpha}\left(S^t_{\sigma\alpha}(p),\ S^t_{\sigma\alpha},\ \neg p;x\right)}{C^t_\alpha\left(S^t_{\sigma\alpha}(p)\right)}$$

$$= \frac{\kappa^t_{\sigma\alpha}\left(S^t_{\sigma\alpha}(p) \mid S^t_{\sigma\alpha},p;x\right) \times \kappa^t_{\sigma\alpha}\left(S^t_{\sigma\alpha},\ p;x\right) + \kappa^t_{\sigma\alpha}\left(S^t_{\sigma\alpha}(p) \mid S^t_{\sigma\alpha},\neg p;x\right) \times \kappa^t_{\sigma\alpha}\left(S^t_{\sigma\alpha},\ \neg p;\ x\right)}{C^t_\alpha\left(S^t_{\sigma\alpha}(p)\right)}$$

Now apply PP_{lim} and CI_{lim} to the factors in both terms of the numerator, and then again the equivalence $S^t_{\sigma\alpha} \wedge S^t_{\sigma\alpha}(p) \equiv S^t_{\sigma\alpha}(p)$:

$$\kappa^t_{\sigma\alpha}\left(x|\ S^t_{\sigma\alpha}(p)\right) = \tau^t_{\sigma\alpha}(x) \times C^t_\alpha\left(S^t_{\sigma\alpha}\right) \times \frac{x \times C^t_\alpha(p) + \bar{x} \times C^t_\alpha(\neg p)}{C^t_\alpha\left(S^t_{\sigma\alpha}(p)\right)}$$

$$= \tau^t_{\sigma\alpha}(x) \times \frac{x \times C^t_\alpha(p) + \bar{x} \times C^t_\alpha(\neg p)}{C^t_\alpha\left(S^t_{\sigma\alpha}(p) \mid S^t_{\sigma\alpha}\right)}$$

We can calculate the denominator in this expression by using the definition of conditional probability and expanding twice using the law of total probability (once using the discrete version, and once using the continuous one):

$$C^t_\alpha\left(S^t_{\sigma\alpha}(p)|\ S^t_{\sigma\alpha}\right) = \frac{C^t_\alpha(S^t_{\sigma\alpha}(p),\ S^t_{\sigma\alpha})}{C^t_\alpha(S^t_{\sigma\alpha})} = \frac{C^t_\alpha\left(S^t_{\sigma\alpha}(p),\ S^t_{\sigma\alpha},\ p\right) + C^t_\alpha\left(S^t_{\sigma\alpha}(p),\ S^t_{\sigma\alpha}\neg p\right)}{C^t_\alpha(S^t_{\sigma\alpha})}$$

$$= \frac{1}{C^t_\alpha(S^t_{\sigma\alpha})} \times \int_0^1 \kappa^t_{\sigma\alpha}\left(S^t_{\sigma\alpha}(p),\ S^t_{\sigma\alpha},p;x\right)dx + \frac{1}{C^t_\alpha(S^t_{\sigma\alpha})} \times \int_0^1 \kappa^t_{\sigma\alpha}\left(S^t_{\sigma\alpha}(p),\ S^t_{\sigma\alpha},\neg p;x\right)dx$$

$$= \frac{1}{C^t_\alpha\left(S^t_{\sigma\alpha}\right)} \times \int_0^1 \kappa^t_{\sigma\alpha}\left(S^t_{\sigma\alpha}(p) \mid S^t_{\sigma\alpha},p;x\right) \times \kappa^t_{\sigma\alpha}\left(S^t_{\sigma\alpha},p;x\right)\ dx$$

$$+ \frac{1}{C^t_{á}(S^t_{\sigma\alpha})} \times \int_0^1 \kappa^t_{\sigma\alpha}\left(S^t_{\sigma\alpha}(p) \mid S^t_{\sigma\alpha},\neg p;x\right) \times \kappa^t_{\sigma\alpha}\left(S^t_{\sigma\alpha},\neg p;x\right)\ dx.$$

Let us refer to the last expression as ψ. Applying CI_{lim}, then canceling, and applying PP_{lim}, we get

$$\begin{aligned}\psi &= C_{\alpha}^{t}(p) \times \int_{0}^{1} \kappa_{\sigma\alpha}^{t}\left(S_{\sigma\alpha}^{t}(p) \mid S_{\sigma\alpha}^{t}, p; x\right) \times \kappa_{\sigma\alpha}^{t}(x)\, dx \\ &+ C_{\alpha}^{t}(p) \times \int_{0}^{1} \kappa_{\sigma\alpha}^{t}\left(S_{\sigma\alpha}^{t}(p) \mid S_{\sigma\alpha}^{t}, \neg p; x\right) \times \kappa_{\sigma\alpha}^{t}(x)\, dx \\ &= C_{\alpha}^{t}(p) \times \int_{0}^{1} x \times \tau_{\sigma\alpha}^{t}(x)\, dx + C_{\alpha}^{t}(\neg p) \times \int_{0}^{1} \overline{x} \times \tau_{\sigma\alpha}^{t}(x)\, dx \\ &= C_{\alpha}^{t}(p) \times E[\tau_{\sigma\alpha}^{t}] + C_{\alpha}^{t}(\neg p) \times E[\overline{\tau_{\sigma\alpha}^{t}}],\end{aligned}$$

Putting it all together, we finally arrive at the result:

$$\tau_{\sigma\alpha}^{t+1}(x) = \tau_{\sigma\alpha}^{t}(x) \times \frac{x \times C_{\alpha}^{t}(p) + \overline{x} \times C_{\alpha}^{t}(\neg p)}{C_{\alpha}^{t}(p) \times E\left[\tau_{\sigma\alpha}^{t}\right] + C_{\alpha}^{t}(\neg p) \times E\left[\overline{\tau_{\sigma\alpha}^{t}}\right]}.$$

References

Bala, V., and Goyal, S. (1998), "Learning from neighbours," *Review of Economic Studies Limited* 65: 595–621.

Barabási, A.-L., and Albert, R. (1999), "Emergence of scaling in random networks," *Science*, 286: 509–512.

Collins, P. J., Hahn, U., von Gerber, Y., and Olsson, E. J. (2015), "The bi-directional relationship between source characteristics and message content," *Proceedings of the Annual Meeting of the Cognitive Science Society* (CogSci 2015).

Douven, I., and Kelp, C. (2011), "Truth approximation, social epistemology, and opinion dynamics," *Erkenntnis* 75(2): 271–283.

Goldman, A. I. (1986), *Epistemology and Cognition*. Harvard University Press.

Goldman, A. I. (1999), *Knowledge in a Social World*. Clarendon Press.

Goodin, R. E. (2003), *Reflective Democracy*, Oxford University Press.

Hegselmann, R., and Krause, U. (2006), "Truth and cognitive division of labour: First steps towards a computer-aided social epistemology," *Journal of Artificial Societies and Social Simulation* 9(3) (available at http://jasss.soc.surrey.ac.uk/9/3/10.html).

Jeffrey, R. C. (1983), *The Logic of Decision*, 2nd ed., University of Chicago Press.

Lazer, D., and Friedman, A. (2007), "The network structure of exploration and exploitation," Computer and Information Science Faculty Publications. Paper 1. http://hdl.handle.net/2047/d20000313.

Lewis, D. (1980), "A subjectivist's guide to objective chance," in Richard C. Jeffrey (ed.), *Studies in Inductive Logic and Probability*, vol. 2, University of California Press.

Mason, W. A., Jones, A., and R. L. Goldstone (2008), "Propagation of innovations in networked groups," *Journal of Experimental Psychology: General* 137(3): 422–433.

Masterton, G. (2014), "Topological variability of collectives and its import for social epistemology," *Synthese* 191: 2433–2443.

Masterton, G., and Olsson, E. J. (2013), "Argumentation and belief updating in social networks: A Bayesian approach," in Fermé, E., Gabbay, D., and Simari, G. (eds.), *Trends in Belief Revision and Argumentation Dynamics*, College Publications.

Miller, D. (1966), "A paradox of information," *British Journal for the Philosophy of Science*, 17(1): 59–61.

Newman, M. E. J. (2001), "The structure of scientific collaboration networks," *Proceedings of the National Academy of Sciences USA*, 98(2): 404–409.

Olsson, E. J. (2011), "A simulation approach to veritistic social epistemology," *Episteme* 8(2): 127–143.

Olsson, E. J. (2013), "A Bayesian simulation model of group deliberation and polarization," in Zenker, F. (ed.), *Bayesian Argumentation*, Synthese Library, Springer, 113–134.

Olsson, E. J., and Vallinder, A. (2013), "Norms of assertion and communication in social networks," *Synthese* 190: 1437–1454.

Redner, S. (1998), "How popular is your paper? An empirical study of the citation distribution," *European Physical Journal B*, 4: 131–134.

Vallinder, A., and Olsson, E. J. (2013), "Do computer simulations support the argument from disagreement?," *Synthese* 190(8): 1437–1454.

Vallinder, A., and Olsson, E. J. (2014), "Trust and the value of overconfidence: A Bayesian perspective on social network communication," *Synthese* 191: 1991–2007.

Zollman, K. J. (2007), "The communication structure of epistemic communities," *Philosophy of Science* 74(5): 574–587.

PART TWO

FORMING COLLABORATIONS

3

Learning to Collaborate

KEVIN J. S. ZOLLMAN

Collaboration involves tackling problems together. Different individuals might bring diverse perspectives to a problem, and by working together they come to solution that none would have reached alone. Collaborating comes with a cost, however. One must expend effort communicate one's approach to another. Collaborations require agreement about the strategies for tackling the problem, and one's collaborator might be difficult or unhelpful. As groups grow the possibility of other epistemic pathologies like group-think and collective ignorance arise.

Whether collaboration is helpful or harmful depends on how these costs and benefits are weighed. Much of the research on collaboration is focused on enumerating, theorizing, and comparing these various costs and benefits. Scholars normally focus on the benefits or harms that collaboration has on those directly involved in the collaborative effort (cf. Kerr and Tindale, 2004). But collaborations also create what economists call "externalities" for those outside of the collaboration. By collaborating with me, you work less with others. If I collaborate with you, you come to learn (at least in part) how I view the world. You might then share this — for better or worse — with future collaborators.

Dealing with externalities can be a tricky matter. Externalities create difficult social dilemmas, like the prisoner's dilemma and tragedy of the commons, where the best thing for the group is inconsistent with individuals' self interest. Even in situations where optimality is consistent with individual choice, externalities can make achieving this good outcome complicated or nearly impossible.

In this article, I utilize a rudimentary model to determine what types of collaborative exchanges would be optimal and under what conditions we should expect groups of scientists to endogenously form optimal collaborative groups. While this model does not represent every collaborative exchange, but it does represent some important aspects of collaboration. By analyzing this model, I find a few lessons for those who wish to maximize the benefits of collaboration. First, by reducing the cost borne by scientists from collaborating, one benefits the members of the group, albeit in a particular way. Second, enlarging the group of potential collaborative partners benefits all those involved in a collaborative group. Finally, and perhaps most counter intuitively, groups do better when scientists have high inertia in choosing collaborative

Boyer-Kassem, Mayo-Wilson, & Weisberg. *Scientific Collaboration and Collective Knowledge*.

partners — that is, groups do better when individual scientists don't try out new collaborative interactions too often.

The model presented in this paper fits within a larger literature of epistemic network models (for an overview, see Zollman, 2013). Collaboration is modeled as creating a social network. This model, like others in this literature, are highly idealized. As a result, the results are suggestive rather than definitive.

3.1 Modeling Collaboration

While collaboration has many facets, the model I present focuses on collaborations with a one-way exchange of assistance. You go to collaborate with me. I help you solve one of your problems, but you don't help me solve any of mine. One can model other forms of collaboration as well. For instance, both Jackson and Wolinsky (1996) and Bala and Goyal (2000) develop models of collaborations that feature two-way collaboration.

The interaction I model is minimally collaboration. The exchange of information and assistance goes only one-way. As a result, this model lies at the border of what we might call "collaboration" and "consultation." By focusing on one-way collaboration, I do not mean to imply other forms of collaboration are non-existent or uninteresting. But already in this limited situation several complexities arise.

In this article I will combine, re-interpret, and extend two different models developed by Bala and Goyal (2000) and Hong and Page (2004). We begin with a fixed group of *n actors*. These actors represent any single individual or cooperatively acting group that is attempting to solve a problem. Actors might be an individual scientist interacting with other individual scientists. Or actors might be a scientific lab interacting with other scientific labs. The model need not be restricted to science, many other groups will fit the assumptions. Henceforth, I will refer to the individuals as "scientists," although the reader should remember that the interpretation of the model can be broad.

Each scientist faces a different problem with many solutions. This might be a high-level problem, like developing a novel theory or designing a complex experiment. The problem might be more simple and mundane, like attempting to prove a small theorem or fix a broken piece of machinery. Whatever its level of significance, the problem must have many solutions that range in quality from worthless to exceptional.

Although the problems faced by the scientists differ from one another, we will suppose they occupy the same field — each scientist can provide assistance on every other problem. Although they work in the same area, each scientist approaches problems in her own way. Each has a unique organization of the space of solutions. The scientists begin with a potential solution to the problem and then move to other solutions that occur to them if those solutions appear better.

Suppose, for instance, an ecologist — call him Carlos — confronts an odd behavior in the field. Carlos finds in the organism he studies, males exhibit bright colors. He wants to figure out which of the many potential explanations for this behavior is the correct one (cf. Maynard Smith and Harper, 2003). Potential experiments abound, and Carlos must try to figure out the one that would be best for uncovering this phenomena.

Carlos begins by considering experimental design *u*. He forms a judgment — I'll suppose a correct one — about how effective design *u* is likely to be at discovering the underlying mechanism of interest. While considering design *u*, Carlos can also imagine designs *w* and *x* as possible. Another ecologist, call her Julie, might have a different conceptual organization of the space of possible experiments. When Julie begins by considering *u* she conceives of designs *y* and *z* as possible. Carlos, if left to work on the problem by himself, would overlook *y* and *z*, while Julie would miss *w* and *x*.

Collaboration in this model is represented by scientists teaching one another their conceptual schemes. So, if Carlos goes to Julie for help, she can provide at least some information about how she conceives of the problem that might bring designs *y* and/or *z* to Carlos' attention.

In this model, gaining access to every additional conceptual scheme is as good as the last — there is no decreasing marginal returns. To put it another way: learning two conceptual schemes is twice as good as learning one, and learning three is three times as good as one, and so on. Finally, each person's conceptual scheme is as good as any other's — everyone is equally smart. This allows me to represent the benefit of collaboration with a variable, b. Gaining access to one collaborative scheme improves one's ability to solve the problem by degree b, two schemes are worth $2b$, and so on. For simplicity I will normalize the expected quality of one's solution without collaboration to zero and normalize the benefit for each additional collaboration b to 1.

But, as we all know, collaboration comes at a cost. Learning another's conceptual scheme takes time; time that might be better spent working in isolation. Worse yet, the new conceptual scheme might be misleading causing one to waste time on inferior solutions. Whatever the source, I will represent this cost by c. We henceforth assume that there is always a cost, by stipulating that $c > 0$.

Each scientist chooses a group of collaborators. Collaborators can, with high fidelity, communicate the relevant parts of all the conceptual schemes they know — both their own and schemes gleaned from others. So, if Carlos collaborates with Julie and Julie collaborates with Shannon, Carlos gains access to both the relevant parts of Julie's conceptual scheme and also to the relevant parts of Shannon's. As mentioned before, all exchanges are one-way. So, if Carlos contacts Julie he gains Julie's conceptual scheme (at a cost to Carlos but at no cost to Julie[1]), but Julie does not gain access to Carlos'.

The model proceeds in two distinct steps. First, everyone chooses which scientists they will contact. Second, schemes are exchanged. This removes complex order effects where it matters whether I contact you before or after you contact another scientist. While this reduces the realism of the model, it greatly increases its tractability.

[1] One might balk at this assumption, since communicating one's conceptual scheme to another might take time. Since Julie does not gain from her interaction with Carlos, Julie would have no reason collaborate with Carlos if it costs her. One might assume, for instance, that part of the cost imposed on Carlos (captured by the variable c) is the amount of remuneration he must give Julie for her time.

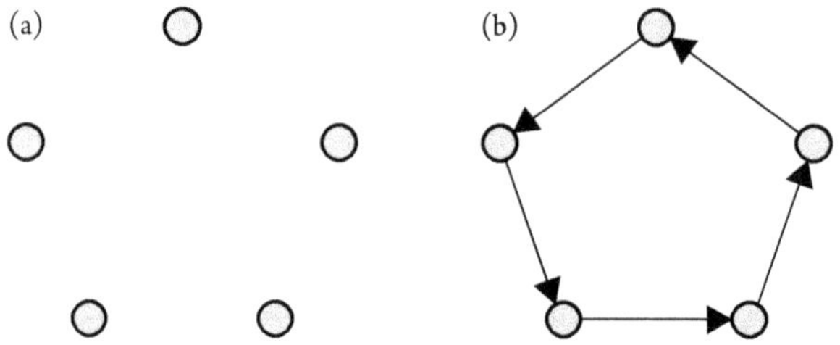

Figure 3.1 Two graphs: the empty graph (a) and the directed cycle (b).

With these assumptions one can now use pictures to represent the outcome of the choices of each scientist (Figure 3.1). Scientist *i* forming a connection to scientist *j* is represented by an arrow from *j* to *i* (representing the direction of the flow of conceptual schemes). A scientist can then be assigned a "payoff" from his choice : the total number of upstream scientists minus the total cost of the connections he forms.

This representation makes clear why models of this form are called "network formation" models. Each scientist chooses whom to link to in a network, and she receives a payoff determined by the entire network's structure. These models are used to represent a multitude of relationships including co-authorship and friendship (Jackson, 2008; Goyal, 2007). But here I will focus on the epistemic interpretation (Zollman, 2013).

There are three distinct metrics for the effectiveness of a collaborative network. The first is called *social optimality*. A pattern of collaboration is socially optimal if it maximizes the sum of the individual payoffs. This represents the situation where all problems are solved to the greatest degree possible given the conceptual schemes available in the community. In this model, only two different states could be socially optimal. When $c \leq n - 1$, collaboration is worthwhile so long as doing so will give one access to a rich set of conceptual schemes. In this context the *directed cycle* is optimal (see figure 3.1(b)). If $c > n - 1$, then collaboration is not worthwhile, and the only optimal state is one where no one collaborates with anyone (Bala and Goyal, 2000).

This result has important implications. Consider for instance the apparently best case scenario for collaboration, $1 > c$. Here collaboration is valuable even if I only gain access to one additional conceptual scheme. In a large group, one might be inclined to infer that everyone ought to collaborate with everyone. After all, if there were a field comprising only two scientists, they ought to collaborate. But, when conceptual schemes are transmitted second hand, this is not true; the optimal structure of collaborative exchanges is sparse. The system would not be improved by encouraging further collaboration.

The reason for this, somewhat counter intuitive conclusion comes from the presence of externalities mentioned at the outset. If Julie has contacted Shannon and learned her conceptual scheme, then Julie is more valuable as a collaborator for Carlos. If Julie can transmit Shannon's scheme with high fidelity, then Carlos does best by working with Julie and gaining access to two additional conceptual schemes. Carlos has little to gain by also interacting with Shannon, since Julie has given him all the information that Shannon might provide. This will hold true even when second-hand transmission of information loses fidelity — up to a point, see Bala and Goyal 2000.

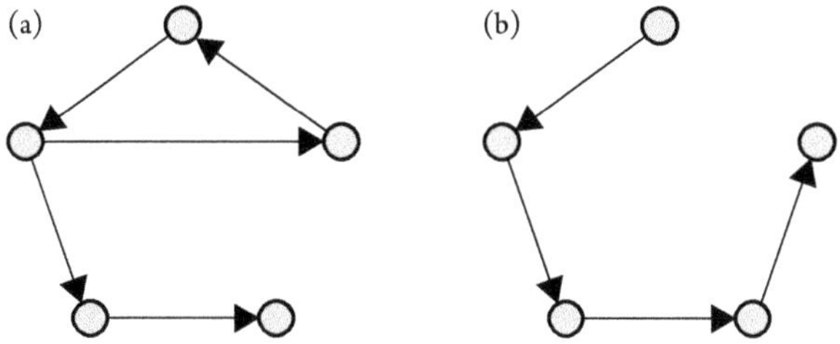

Figure 3.2 Two unstable graphs.

Social optimality in games is a technical concept which imperfectly approximates our intuitive notion of "good for the group." First, if one is modeling scientists researching a problem that has implications for those outside of science, one also might want to consider the impact of the research on society at large. Under the intended interpretation of the model provided here, the payoff to scientists is just the degree to which they solve the problem. If each scientific problem is valuable (and equally valuable) to the world-at-large, then what is socially optimal for the community of scientists reflects the good of the community at large.

One might worry about social optimality because one objects to maximizing the sum or average payoff. This focus is reasonable for groups of scientists since we are primarily interested — as a society — in the total progress of science. In many political settings, in contrast, some advocate for giving special consideration to the worst off. Egalitarian considerations reach the same conclusion in this model since the worst off does best in the state I call "socially optimal." Other measures are possible, too, and some of them will not so neatly coincide with my choice for a measure of group effectiveness. However, given the intended interpretation of the model, they cannot be given much defense.

The second metric of evaluation for a pattern of collaboration is stability. A pattern of collaboration is stable, if no individual scientist is able improve his situation by changing his collaborative partners. For instance, the pattern pictured in figure 3.2(a) is not stable, because the individual in the upper right can do better by connecting to the individual in the bottom right. By doing so, she would increase her payoff from $2 - c$ to $4 - c$.

In this model, socially optimal states are always stable (Bala and Goyal, 2000). When $c < 1$ scientists are willing to collaborate with one other scientist even if the potential collaborator can only provide her with one new conceptual scheme. In these cases, only the directed cycle is stable — in any other situation at least one scientist has a positive incentive to change her pattern of collaboration.[2] When $n - 1 > c > 1$, the directed cycle remains stable but the state where no one interacts is also stable (see figure 3.1(a)). Here collaboration cannot get "off the ground" because no one is willing to take the first step. The first person to collaborate must pay the cost to gain access to one conceptual scheme, and since c is greater than the value of a single scheme, no one wants to pay the cost. We have an example of a sub-optimal, but stable,

[2] When $c = 1$ things are slightly more complex. A scientist has neither a positive nor a negative incentive to change from some other patterns of interaction.

outcome — a situation familiar to many of us. Turning to the last case, when $c > n - 1$, we find the only stable state is the one where no scientist collaborates with any other, which is also optimal.

The last metric of evaluation, which will be our primary focus, is one of "learnability." When $n - 1 > c > 1$, there is more than one stable state. Even if scientists always find a stable states, one cannot be assured that they will come to land on the socially optimal one. Scientists might not come to find a stable state, but wander around moving from one unstable state to another, perhaps indefinitely. To uncover which states are learnable, one must model of learning.

3.2 Modeling Learning

In order to model learning, I now must extend the model of one-shot collaboration to repeated collaboration. Now scientists are confronted with problems sequentially. Each time a scientist is confronted by a new problem, she can connect to others and collaborate on *that* problem according to the model above. Each new problem differs from the previous one, and each instance of collaboration only provides enough information about a conceptual scheme to solve the problem under consideration. This assumption seems plausible as collaboration is usually problem-focused. The resulting model is tractable because the benefits from each instance of collaboration is independent of the previous collaborations.

Many models of learning in game theory require scientists to form a belief about what others are doing. In games of this type, this requirement amounts to having a probability distribution over all directed graphs with n nodes — a large space even for moderately sized n. Not only would this be difficult to analyze, it is unlikely to provide any real insight into how scientists behave.

Instead Huttegger et al. (2014) suggest this game is best analyzed using a learning rule called "probe and adjust" which represents unadorned learning. Each scientist has a default strategy which she usually employs. Occasionally, a scientist experiments – she tries a new strategy at random (a probe). If the new strategy outperforms the default, she adopts it as her new default. On the other hand, if the new strategy is inferior to the old one, she returns to the previous default. If the two are tied, she chooses which one will be the new default at random.[3]

This method of strategy revision has a few interesting properties. If one enforced the rule that there must be one round of default play in between single probe events, then the process would be guaranteed to evolve into a stable state and then the default behavior would never change. While mathematically helpful, limiting the system in this way seems implausible — how would scientists ensure they weren't probing

[3] Whether or not probe and adjust represents *exactly* how scientists are learning is not important. The tractability of probe and adjust makes this model an important starting point from which more general lessons can be drawn. The phenomena illustrated in the next section strikes me as sufficiently general as to reoccur in almost any learning rule where scientists are unaware of the other connections in the network — that is, where scientists do not know the patterns of collaboration.

simultaneously or immediately after another? If one relaxes this assumption the group can escape from stable states, even optimal ones.

As an example, consider the directed cycle (figure 3.1(b), the optimal state), where $c = 3$. Suppose the individual in the upper right probes by connecting to the individual in the upper left (figure 3.2(a)). Not only does this lower her own payoff, but her probe lowers the payoff of everyone else except for the person on the lower right. Of particular importance for this example is the individual on the top, whose payoff is now $2 - c = -1$. After one round of probing, the individual on the top left will switch back to her default strategy. If the individual on the top probes on the subsequent round by abandoning collaboration altogether, his situation has improved over the previous round — his payoff was -1 (because the other individual was probing) and now it is 0. So, he will stick with this strategy (figure 3.2(b)). Now the individual on the upper left will prefer to drop his connection, and so on. The system has now abandoned an optimal state.

3.3 Results

The overarching question is, to what extent can scientists come to collaborate efficiently? This is a multifaceted question that is addressed in parts.

First, to what degree is experimentation with collaboration patterns helpful? Is the group improved if people are constantly exploring — perpetually probing — or should they be more set in their ways? The answer to this question is not obvious. On the one hand, high probing rates will help the system escape sub-optimal ones relatively quickly. But, high rates will also increase the probably that the system will leave an optimal state in favor of a sub-optimal one.

A second research question concerns how the cost of collaboration, c, affects the system. If every group always attained optimality, then everyone improves as c becomes smaller. But, if we cannot be assured of optimality, the answer is not so straight forward. While a lower c will make the final payoff in the higher for everyone, it also provides a smaller incentive to find the optimal state. A lower c might cause the community to wander around more. Furthermore, lowering c might also affect the ease with which the system escapes the optimal state.

The last research question relates to group size. Smaller groups are likely to reach optimality more quickly. As Huttegger et al. (2014) point out, the number of non-optimal states grows quickly as new scientists are added, but the set of optimal states grows much more slowly. So, the proportion of states that are optimal approaches zero as the number of scientists approaches infinity. On the other hand, the payoff for scientists *even in non-optimal states* goes up as the set of potential collaborators increases (assuming we hold c constant). How do these two considerations trade off against one another?

3.3.1 Experimentation rates

Huttegger et al. (2014) prove that when $c < 1$ (i.e., when the only stable state is also the only optimal one), the system will spend most of its time in the optimal state —

the directed cycle — in the limit. Stated more formally, if you choose a probability p other than 1, there is a probe probability where the long-run probability the system is in the optimal state is at least as high as p. This proof is achieved by showing that from any state there is a chain of intermediate states where (a) at each stage only one player must change to go from one state to the next, (b) the person who changes does not make herself worse off, and (c) the end of the chain is the directed cycle.

This result is limited. First, we cannot generalize to cases where $c > 1$ — the theorem is clearly no longer true. Second, the result does not systematically answer the question about experimentation rates. Finally, we only know what happens in the infinite limit which can be very different from short-run results.

In order to provide a more general analysis, a simulation study was conducted. Simulations considered group sizes ranging from 4 to 10 with probe probabilities ranging from 0.01 to 0.21 in 0.03 increments. c varied from 0.1 to $n + 0.1$ in 0.5 increments.[4] For each setting of the parameters 1,000 simulations were generated where the probe and adjust process continued for 100,000 rounds.[5]

Figure 3.3 illustrates the conclusions regarding probe probabilities. The top plot shows how well scientists fared when measured by the time they spent in the optimal state.[6] Non-optimal states can nonetheless be better or worse. The bottom plot shows how well the groups fared when measured by normalized payoff (0 is the worst possible payoff for that configuration and 1 is the best possible payoff). Both plots reveal the same general trend: lower probe probabilities are better. (This generalizes simulation results presented in Huttegger et al. 2014 for a small cost version of this game.)

Obviously a probe probability of zero would be inferior because the group would never change. However, low probe probabilities do much better than higher ones. Of those tested, a probe probability of 1% was superior. In communities appropriately modeled in this way, one should be cautious about encouraging scientists to experiment with new patterns of collaboration. These simulations suggest that encouraging exploration will hinder the ability of the system to find and remain in the optimal state. Beyond that, increased experimentation will harm the communities ability to do well even when not in an optimal state.

3.3.2 Cost

What effect will the cost of collaboration have on the overall performance of the system? It seems intuitive to strive to minimize the cost of collaboration, but will

[4] The results for these group sizes and probe probabilities show clear patterns that can be projected to larger group sizes and probe probabilities (but not without some care to *smaller* probe probabilities).

[5] 100,000 rounds were chosen in order to model a long but finite time. This will provide a helpful contrast to the limiting results of Huttegger et al. (2014). The results between the limit analysis and this, very long, time differ in significant respects. Since real scientists are operating at much shorter time spans, there might be further differences. The fundamental differences between finite and infinite limit analysis can be identified at this long time scale. This satisfies the duel goals of understanding the system while not risking a time scale that is "too short."

[6] In general there is also a strong relationship between time spent in the optimal state and the first time optimality is obtained.

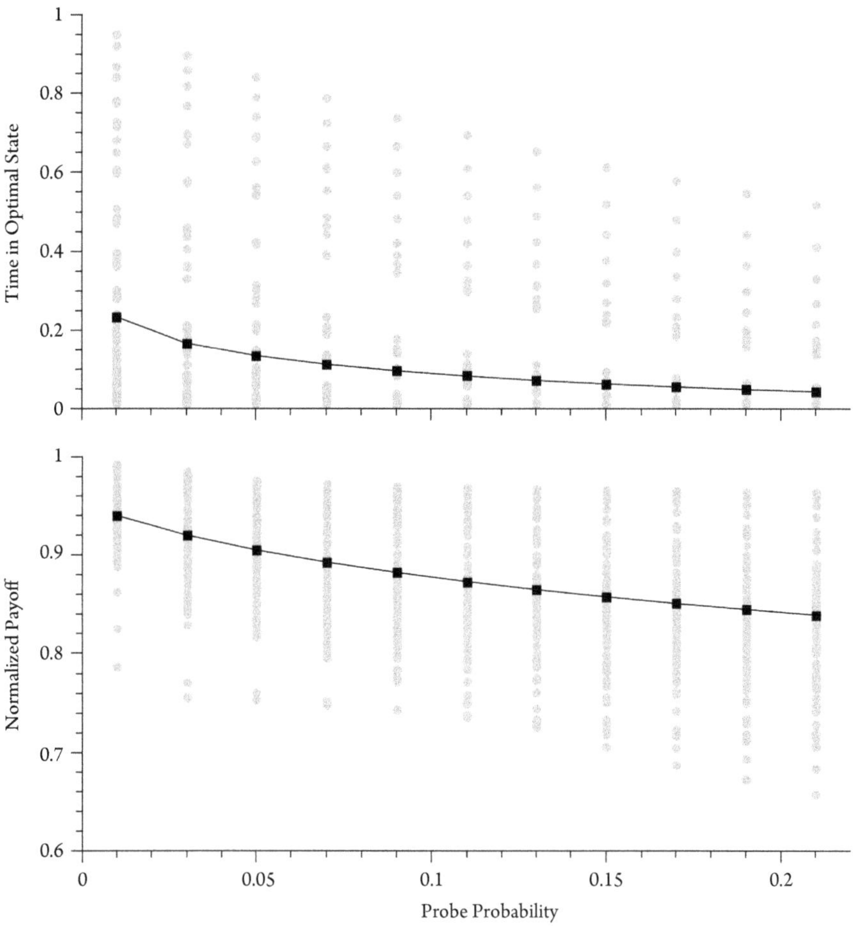

Figure 3.3 Two plots which relate the probability of a scientist probing to the proportion of time spent in the optimal state (top plot) and to the average normalize payoff (bottom plot). In both plots the gray circles represent the average of 1,000 simulations for each of the parameter settings. The black squares represent the average of all simulations for all parameter settings which use the same probe probability.

the simulations endorse this policy? To answer this question I will focus on a representative group size, 7 individuals, all of whom have a low probe probability of 0.01. All the qualitative facts reported here are true of the other group sizes studied.

First let us consider how cost relates to the ability of scientists to find optimality. This is pictured in the top plot of figure 3.4. Recall the three different regimes. When $c < 1$, the optimal state is the only stable one. Here, however, the system is rarely in the stable state. This is consistent with results reported by Huttegger et al. (2014). They conjecture that it is difficult to find the optimal state because the search space is so large. Indeed, this is likely part of the problem. But the size of the search space

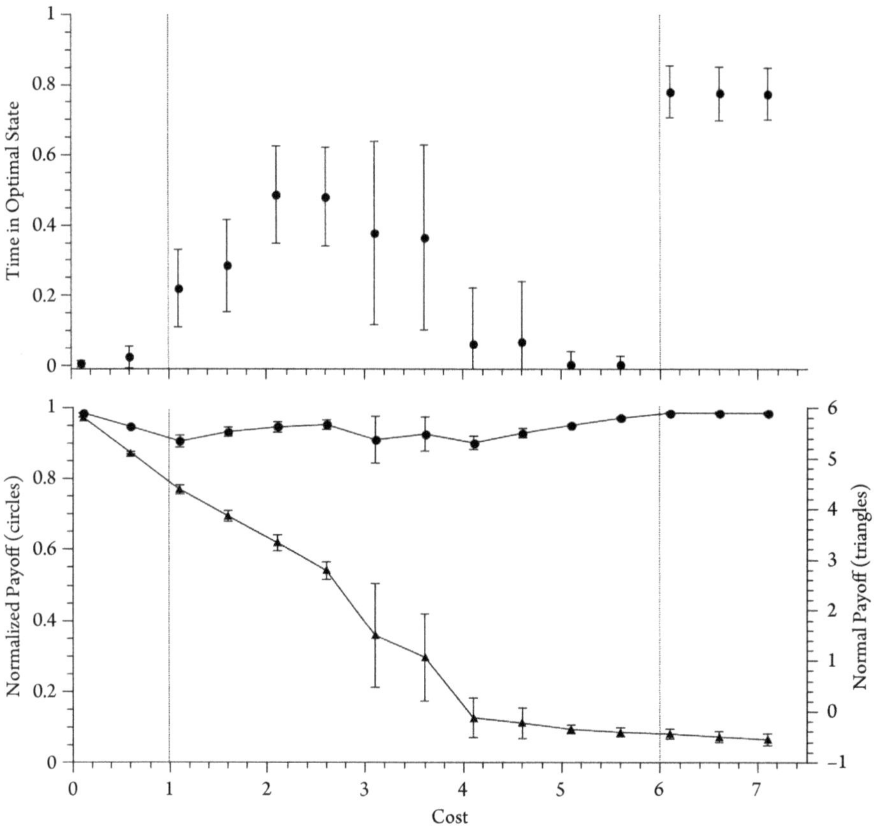

Figure 3.4 Simulations results for a seven-person group with probe probability set at 0.01. The plots represent the average payoff and standard deviation for different values of cost versus time in the optimal state (top plot) and normalized payoff (bottom plot, circles plotted against the left y-axis) and nominal payoff (bottom plot, triangles plotted against the right y-axis).

cannot be the whole story, because when the cost is higher in this region ($c = 0.6$) the group performs slightly better than when it is lower ($c = 0.1$). Furthermore, comparing the low cost regime with cases where $c > 1$ suggest the situation is rather more complicated.

Turning to the second regime, where $6 > c > 1$, we find a complicated relationship between cost and time in the optimal state. There is an optimal cost between 2 and 3 where the system is best (in this regime) at finding, and remaining in, the optimal state. While the changes in cost do not effect what state is optimal, they do affect the ordering of non-optimal states, and in so doing make the optimal state easier to find. Considering only those simulations that occupied the optimal state for at least one round, the system with a cost of 2.1 was over three times faster at finding the optimal state than the system with a cost of 0.1.

One should be careful about jumping to the conclusion that one should attempt to increase the cost of collaboration. While the identity of the optimal state is not changed as the cost increases, the objective quality of this state changes. When the cost is 0.1, each individual in a seven-person group receives a payoff of 5.9 in the optimal state, but when the cost is 2.1, each individual receives a payoff of 3.9. So while the system spends more time in the optimal state, this is not equivalent to improving how individuals fare.

And in fact, they are not faring better with higher costs. The bottom plot of figure 3.4 shows both the normalized payoff and the nominal payoff. The later is most important. While increasing the cost increases the time spent in the optimal state, doing so also makes that state worse. The later consideration outweighs the former. While the low cost communities wander around for a significant amount of their time, they nonetheless do well.

Finally, in third regime, where $c > n-1$, scientists are adept at achieving optimality. This is not surprising, because in this situation connecting to no one dominates connecting to anyone. Because connecting to no one is one strategy out of 64, individuals must find the dominant strategy which is why they are not in the optimal state the entire time.

These results can underwrite the intuition that collaboration is improved by making collaboration less costly (by perhaps improving mechanisms of communication, or by providing direct incentives designed to compensate for other costs). But, reducing the cost does not increase the chance that the community arranges itself in optimal ways — to the contrary.

3.3.3 Community Size

As community size increases, it will become more difficult for the group to find the optimal state (Huttegger et al., 2014). But, as in the previous section, increasing the population of scientists also improves the nominal payoff of that best case. It therefore is an open question whether increasing the size of the community will be beneficial or harmful to that community.

Figure 3.5 illustrates the settings of the parameters which maximize the time in optimal state for each group size. While, as expected, the time spent in the optimal state decreases rapidly as the group size grows, the benefit from the larger size swamps the loss. The nominal payoff increases despite a greater fraction of the time is spent in non-optimal states.

There is a harm to increasing the group size ; they will spend more time exploring non-optimal states. In this model, that harm is worth incurring because the benefit of increased collaborative possibilities is sufficiently large to outweigh the cost (at least in the best case).

3.4 Conclusion

Not all collaborative interactions are analogous to this model. I have argued that there are unlikely to be any fully general results about the structure of epistemic networks

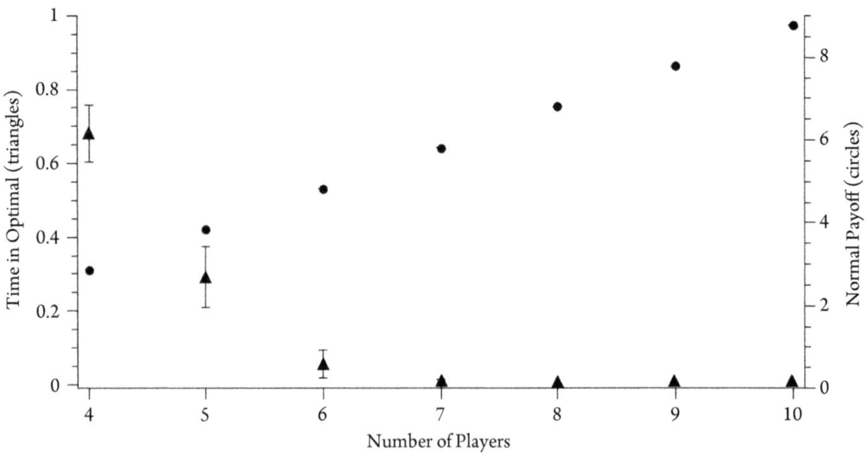

Figure 3.5 Simulation results comparing the number of players to the time in the optimal state (triangles plotted against left y-axis) and to the nominal payoff (circles plotted against the right y-axis). Each point represents the single probe probability and cost that maximized the nominal payoff relative to other values of those parameters for groups of the same size.

(Zollman, 2013), and I expect this will be true with collaboration as well. However, I do believe that this model provides an appropriate idealization of some situations of collaboration, and where it does, it offers some clear guidance to on how to best achieve effective collaboration.

The model underwrites the general belief that productive collaboration is facilitated by (a) increasing the number of individuals with whom collaboration is possible and (b) decreasing the cost to scientists for engaging in collaborative exchanges. These are true despite a reduction in the time spent in optimal states.

More surprisingly, however, the model suggests that encouraging scientists to experiment more with different collaborative arrangements will not be productive. With the increased interest in interdisciplinarity, funding agencies and administrators have been encouraging scientists to find new patterns of collaboration. However, in this model, spuring new collaboration by introducing new scientists to one another or by encouraging scientists to try something new, is counterproductive.

References

Bala, V. and S. Goyal (2000, September). A noncooperative model of network formation. *Econometrica 68*(5), 1129–1181.

Goyal, S. (2007). *Connections: An introduction to the economics of networks*. Princeton: Princeton University Press.

Hong, L. and S. E. Page (2004, November). Groups of diverse problem solvers can outperform groups of high-ability problem solvers. *Proceedings of the National Academy of Sciences of the United States of America 101*(46), 16385–16389.

Huttegger, S., B. Skyrms, and K. J. S. Zollman (2014). Probe and adjust in information transfer games. *Erkenntnis 79*, 835–853.

Jackson, M. O. (2008). *Social and economic networks*. Princeton: Princeton University Press.
Jackson, M. O. and A. Wolinsky (1996). A strategic model of social and economic networks. *Journal of Economic Theory* 74(0108), 44–74.
Kerr, N. L. and R. S. Tindale (2004, January). Group performance and decision making. *Annual Review of Psychology* 55, 623–655.
Maynard Smith, J. and D. Harper (2003). *Animal signals*. Oxford: Oxford University Press.
Zollman, K. J. S. (2013). Network epistemology: Communication in epistemic communities. *Philosophy Compass* 8(1), 15–27.

4

Diversity, Rationality, and the Division of Cognitive Labor

RYAN MULDOON

4.1 Introduction

Why do scientists cooperate? Why do some disciplines, like physics and computer science, appear to be spreading into new domains, like biology? Why is it that scientists appear to be more specialized over time? These are all basic questions about the division of cognitive labor in science. They are also questions that models of the division of cognitive labor in science that have been developed by philosophers simply cannot answer. Despite several different models of the division of cognitive labor,[1] none of them has a structure that enables these questions to be asked of the model. This not only narrows the questions that can be investigated in the philosophy of the social structure of science, it also impedes our ability to understand the division of cognitive labor itself.

Weisberg and Muldoon (2009) argued that the division of cognitive labor is best understood by assuming that scientists are exploring an unknown landscape. In this model, scientists have a particular problem domain in mind and are searching for different approaches to the problem domain. With this setup in mind, Weisberg and Muldoon then fruitfully explore different methods of dividing cognitive labor—how scientists choose new projects based on the work that others have done. In particular, they consider three strategies for responding to the evidence of other scientists' research: they can ignore it altogether (controls), they can follow in others' footsteps (followers), or they can intentionally work in areas where others are not (mavericks). Weisberg and Muldoon show that while taking in evidence of others' work is important, how one does it matters more: a population of followers slightly underperforms a

[1] Muldoon (2013) offers a survey of the prominent modeling approaches to the division of cognitive labor. Most notable are Kitcher (1990, 1993), Strevens (2003), Weisberg and Muldoon (2009), and Zollman (2010). These authors largely aim to stay within Kuhnian normal science. Wray (2014) offers a Kuhnian account of specialization in science. De Langhe (2014) offers a model of how theories themselves can be accounted for by the division of cognitive labor. In this chapter I focus on project-level cooperation and competition, whereas both Wray and De Langhe look at a more macro-level of cooperation on a particular paradigm.

Boyer-Kassem, Mayo-Wilson, & Weisberg. *Scientific Collaboration and Collective Knowledge*.

population of controls, while a population of mavericks massively outperforms either. Further, adding mavericks to a population of followers not only improves overall social outcomes but also improves the epistemic outcomes of the followers themselves. This has proven to be a useful framework for thinking about important epistemic issues in the division of cognitive labor, but it has several shortcomings. In particular, the model assumes that there is no cost to adopting new approaches—skill acquisition or equipment change is simple. Mavericks are an especially clear example of this—they constantly work on novel approaches, even though in the real world this may be rather costly. This is a natural assumption to make when assessing different ways to respond to evidence, and the authors do note that this is an idealization of their model.

This assumption—that scientists can change their skills on a whim—may be reasonable for the narrow domain that Weisberg and Muldoon have for their model, but it is clearly unrealistic as we seek to understand slightly more general features of the scientific community. In real labs, scientists tend to make rather costly upfront investments in skills (graduate school and several postdocs are common for training periods) or equipment (lab equipment can regularly cost hundreds of thousands of dollars, besides the training in using and maintaining the equipment). These costs create quite steep barriers to entry for a lab that wishes to employ skills that they do not currently possess.

In this chapter I argue for a modeling approach that inverts the Weisberg and Muldoon approach. Rather than assume that scientists have more-or-less fixed questions and can easily acquire new skills, I assume that scientists have a fixed set of skills and search for problems that their skills enable them to solve. This does not refute the Weisberg and Muldoon approach—their core model results remain intact—but it offers a more flexible framework with which we can examine scientific communities and offer normative guidance.

An immediate benefit of such an approach is that it helps us address a core area of science that has gone virtually ignored by the major recent philosophical works in the division of cognitive labor (Kitcher 1990, Strevens 2003, Zollman 2010, Weisberg and Muldoon 2009): scientific collaboration. In all the major models of the division of cognitive labor, competition is always better than cooperation.[2] However, we see a great deal of cooperation among scientists. Rather than assume that scientists are massively irrational, if we suppose that scientists are limited in the set of skills that any one of them has, we can see why cooperation should emerge in some cases and not in others. In particular, we should expect cooperation to emerge when scientists work on "harder" problems—problems that require a wider range of skills than one lab typically has. Competition remains where problems are sufficiently simple such that a single lab has the resources to tackle the problem.

By adding extra constraints to models of the division of cognitive labor, it becomes easier, not harder, to generate key features of the social structure of science. This framework then becomes an excellent platform for asking more strictly normative questions about how scientists under constraint can collectively maximize their scientific output.

[2] In this essay, I am focused on project-level competition and cooperation. Wray (2014) and De Langhe (2014) discuss cooperation at the level of paradigm selection.

To proceed, I will first examine the current trends of collaboration in science and the costs that shape the nature of scientific investigation. I will argue that these costs are significant enough to the decision-making of scientists that we must find ways of representing them in our models. Next, I will argue how these constraints enable models to help us investigate two key phenomena: scientific collaboration, and disciplinary expansion. Finally, I will conclude with a discussion of what this modeling approach means for the division of cognitive labor more generally.

4.2 What Features of Science Are We Missing?

While academic science is undoubtedly a highly competitive environment, with heavy pressures to be the first to publish, it is also increasingly a collaborative environment. Glanzel and Schubert point out that since 1980, single-author papers have gone from approximately a quarter of all scientific papers, down to only 10% in 2000. The number of coauthors on papers has also steadily risen, from a mean of 2.64 to a mean of 4.16 in the same time period. The authors argue that "all areas of science are characterized by intensifying collaboration associated with the increase of the share of multi-authored papers" (p. 262). Further, this increased cooperation is associated with increased productivity of individual scientists, though the optimal level of collaboration appears to be different between disciplines—math is at the low end with a peak around two coauthors, whereas chemistry does best with three to four, and neuroscience and biomedical research requiring five or six coauthors to reach maximum author productivity (p. 263).

The National Science Foundation (NSF) has also been tracking the increase of collaborative science. Looking at a 20-year window starting in 1990, the NSF has found that in journals covered by the Science Citation Index, the mean number of coauthors has increased from approximately three to approximately six. They, too, found a range in the number of coauthors between disciplines, finding math and the social sciences the lowest with a mean of approximately 2, and astronomy being the most collaborative field with a mean of 14 authors per paper in 2010 (NSF 2012, fig. 5-24). Not only has the NSF found that coauthorship has increased, they have also found that cross-institutional coauthorship has been on the rise. Between 1990 and 2010, the number of institutions listed on a paper grew at a rate of 4.1% annually, a massive rate of increase. Not only this, but the rate of articles with international collaborations has also grown from 10% in 1990 to 24% in 2010 (NSF 2012, chapter 5). These last figures are particularly striking, as one would expect that collaboration across institutions is more difficult than intra-institutional collaboration, and international collaborations still more difficult. In each case, we should expect that communication costs and related logistical challenges should increase. It's easier to work with someone down the hall than someone in a different city. It is unlikely that such collaborations would be taken on at a whim.

The data suggest that collaborations are becoming increasingly integral to scientific production. In a 30-year period, we have seen (on average) the tripling of the number of coauthors per scientific paper and an increase of scientist productivity correlated with this increase of coauthorship. We also know that since we are finding very large increases in inter-institutional collaboration, that this increase in coauthorship is not merely an artifact of scientific labs getting larger, where a Principle Investigator (PI) simply has more postdocs and graduate students working for her. Instead, we have robust evidence of inter-lab cooperation.

Furthermore, we find that the number of coauthors varies across scientific disciplines, and, at least in some disciplines, this matches optimal productivity levels for scientists.[3] This last point is important, as it suggests that these trends are not merely a shift in academic norms but instead fundamental improvements in the social structure of scientific production. Rather than astronomers being friendlier than mathematicians, it appears to be the case that the problems they are working on simply require more people to make progress.

There are several potential drivers for this change in scientific practice. I will focus on two. First, it is unlikely to be a coincidence that in this 30-year period examined, collaboration technologies advanced extremely rapidly. In 1980, inter-institutional collaboration would have to be conducted via postal mail, expensive long-distance phone calls, and pre-deregulation air travel. In 2010, collaborators can use e-mail and free video calling to communicate and exchange large data sets over the Internet at essentially no cost. Air travel is also significantly less expensive (if less pleasant). So undoubtedly, the time and money costs of inter-institutional collaboration have significantly reduced. If we make the standard economic assumption that people consume more of goods when they become cheaper, then it should be unsurprising to find scientists "consuming" collaborations at higher rates as they become cheaper.

The second potential driver for this change in scientific practice is that there has been 30 years of scientific advancement. That is, the easier, or at least more obvious, theoretical and experimental work is more likely to have been done already. Standard accounts of scientific production as a large-scale competition provide the explanatory framework for such an account. In the rush to build up one's publication record for tenure and to build one's reputation as a good scientist, it makes sense to first focus on those projects that will reliably generate papers within reasonable time frames. As an increasing number of scientists compete for papers with similar incentives, we will find that the low-hanging fruit will be more likely be consumed before the more difficult-to-reach fruits. As this happens, what the community interprets to be the low-hanging fruit will continue to ratchet up in difficulty or complexity. As projects get more difficult, the time pressure to publish before one's competition will encourage scientists to seek out collaborators.

While this may seem like an obvious point—that the problems scientists face increase in average complexity over time—it is rare that the nature or difficulty of problems is included in models of the division of cognitive labor. A world of easy problems is very different than a world of hard problems.[4] While some problems might

[3] Optimality is here understood as maximizing epistemic output. The proxy that Glanzel and Schubert (2005) use is number of peer-reviewed articles published. We could imagine different objective functions that could be applied, such as citation counts or peer reputation, but number of papers published is a reasonable starting place.

[4] As defined before, a problem is harder when it is more complex, which means it requires more skills to solve. This fails to differentiate between an 'easy' problem that can be done quickly and an

become easier over time thanks to scientific and technological advancement, we still face the basic dynamic of a reduced proportion of easy problems to hard ones. This means that contemporary scientists, in the context of their research decisions, face a different world than their forebears, even if we are to ignore any institutional changes.

This second factor is even more complex than initially stated. If it is the case that, over time, scientific advancement becomes more difficult because the remaining problems become more complex, a rational response is to not just increase the number of one's collaborators but also to specialize further. As the problems get more difficult, and science advances, there is simply more to learn, whether it is theoretical tools or experimental techniques or just getting up to date on the findings of the field. This requires a longer training period. The NIH has reported that the average age of a first-time R01 grant principal investigator has gone from approximately 36-years-old in 1980 to approximately 42-years-old for a PhD, or about 44-years-old for an MD or MD-PhD.[5] This is an increase of six to eight years before a scientist is functioning as a head of a lab. Given that training periods (often in the form of multiple postdocs) have already extended to the point of the average scientist not being a head of a lab for more than 20 years after completing her bachelor's degree, the value of specialization is all the more apparent. To have a hope of being a productive autonomous contributor to science prior to retirement, one needs to be at the cutting edge of some (increasingly) narrow aspect of one's field. Generalists in a given field take on a potentially enormous training burden just to be conversant across sub-disciplines, and this will only increase over time as fields progress further.

We find that each of the two main drivers of increased scientific collaboration relate importantly to shifting costs. The first driver is a story of reduced expense of collaboration. The second driver is a rather striking story of the increased cost of discovery. These cost changes together conspire to drive scientists toward increased levels of collaboration.

These drivers of collaboration tell an important story about the shifts that help explain this change of the nature of scientific production in the last 30 years. Yet, they are not at all present in the most prominent models of the division of cognitive labor in the philosophy literature. In the next section, I will turn to the question of whether we can ignore these features of scientific practice.

4.3 Capturing the Division of Cognitive Labor in Models of Science

Leaving collaboration out of models of the division of cognitive labor fails to capture the division of cognitive labor. Kitcher (1990) introduced the idea of a "sullied"

'easy' problem that may take a great deal of time and effort. But this is an idealization I am willing to accept, as it's not clear that there is a systematic way in which those two undifferentiated problems affect the division of labor—some problems may be amenable to parallel work, others may not. It is likely true, however, that as time passes, the share of the quick easy problems has declined, for the same reasons we see fewer easy problems in general.

[5] Rockey (2012).

epistemic agent by showing that scientists who care about non-epistemic rewards, such as fame, are more likely to distribute their efforts efficiently across a scientific community. Strevens (2003) expanded on that model, exploring how different incentive schemes might improve on the basic result that Kitcher presented, to argue that the Priority Rule in science was epistemically optimal. Zollman (2010) explored the benefits of increasing epistemic diversity among agents by making them either more stubborn or less aware of what others worked on. Weisberg and Muldoon (2009) showed how different responses to what others worked on could drive significant changes in scientific output. But in none of these approaches do we find anything other than labs that have equal epistemic strengths competing to produce the best science the most quickly. Nowhere in these models is a division of labor in the way that Adam Smith initially meant the term: where different people specialized at different tasks, each of which could contribute to a larger whole.

There are two ways of thinking about the division of cognitive labor. The first, which has been explored starting with Kitcher, looks at how across all of science, scientists allocate themselves to different projects. In this sense, they divide their labor in the same sort of way that butchers and bakers divide their labor in a market economy. This allows for a diversified market, and this can allow for a diversified scientific community as well. Wray (2014) and De Langhe (2014) both develop an account of specialization and the division of cognitive labor in this more macro sense. However, the second way of thinking about the notion of a division of cognitive labor is much closer to the original instantiation of the notion of a division of labor from classical economics.

Adam Smith's most famous illustration of the division of labor is his discussion of a pin factory. He argued that while it would be difficult for an average laborer to produce a single pin in a day, given his lack of familiarity with the process and the machines involved in its production, the task can be broken up into many different parts. Each worker only needs to learn one single task, like drawing wire or straightening it. Smith estimates that the productive increase from having 10 people work on 10 different specialized tasks in pin production as opposed to each merely trying to make pins wholly on their own hovers around 4,800-fold. The story of the productive increase of the Industrial Revolution is that of moving from craftsmen to divided, specialized labor.[6] This is a story about divided but joint production. The pin factory is further distinguished from a shirt factory where we see a different sort of division of labor among firms in the wider economy, but within the pin factory we see specialization and joint production.

Smith also importantly pointed out that the division of labor is only limited by the extent of the market.[7] That is, the larger the market and the attendant labor force, the more finely we can divide labor. The NSF estimates that in 1950, 182,000 people were working in science and engineering, compared to 5.4 million in 2009. This is a nearly 30-fold increase in the labor supply in the last 60 years. Approximately 41% of those with science and engineering doctorates are employed in academic institutions.[8] This

[6] Smith 2003, Book 1, Chapter 1.

[7] Ibid., Book 1, Chapter 3.

[8] NSF 2012, Chapter 3. Available at http://www.nsf.gov/statistics/seind12/c3/c3h.htm.

suggests that the extent of the market has grown considerably, which would support further division of labor, paired with greater collaboration.

The division of cognitive labor, at least used in the way that Smith originally envisioned, is fundamentally about specialization and collaboration. Yet, none of our models of the division of cognitive labor in science explore this dimension of the structure of scientific production. This is contrasted with Peter Galison's excellent work in *Image and Logic: A Material Culture of Microphysics* (1997), which is a careful examination of real scientific practice. Unsurprisingly, Galison demonstrates the essential need for and challenges of collaboration in the context of a division of cognitive labor.

Galison convincingly demonstrates how crucial the division of cognitive labor is in modern scientific production. Smith points out how the division of labor is a crucial input to productivity growth. Yet our philosophical models of the division of cognitive labor do not include specialization or collaboration.

So, the question that we face is whether this is an acceptable omission. All models are idealized versions of the target systems they represent. That we carry out investigations of idealized representations does not mean we have done anything illicit. In fact, I am not looking to argue that any of the models of the division of cognitive labor I have referenced would have their results reversed if these features were built into their models. All the authors mentioned can represent larger collaborations as single labs and largely sidestep the issue.[9] Their papers remain interesting and important contributions to our understanding of the social structure of science. However, my worry stems from the view of science that has thus far been presented by philosophers is both misleading as a descriptive account of science and cuts off the possibility of investigating crucial questions related to improving the production of science.

Just as Kitcher injected some additional realism into the study of science by making the assumption that scientists ought to be thought of as epistemically sullied in the sense that they care about non-epistemic rewards, we ought to also inject additional realism into our models to take into account the costs that scientists face. Only once we take proper account of these costs can we think more carefully about how scientists face decisions about their research. Once we have a firmer descriptive account of research decisions, we are in a much better position to investigate normative questions related to the social structure of science.

4.4 The Costs of Doing Science

As discussed earlier, the drivers of the empirical increase in collaboration in science appear to be changes in the costs of doing research. In this section, I will more carefully lay out the nature of these costs. We will then be in a position to examine the decision environment of scientists.

[9] As nothing hinges on what counts as an agent in any of these models, we can simply redefine the agent from a scientist to a lab or a collaboration across labs. This doesn't change any of the mechanics of the models, it just tweaks the idealization.

The most basic cost is that of skills acquisition. A principle investigator has already invested in a long training period to acquire the set of skills that she currently possesses. These skills, as discussed previously, are becoming increasingly specialized. Scientists, then, face large training costs for increasing their skill set. Developing knowledge of a particular theoretical or experimental technique represents a large time cost. Not only does it take effort and time to learn new skills, it represents a potentially large opportunity cost. The scientist could have been productively exploiting her own existing skill set instead of expanding it.

A related cost is equipment. In experimental science, bench equipment can cost hundreds of thousands of dollars, and in very large-scale experiments, such as the Large Hadron Collider, the equipment costs run into the billions. Insofar as a change of experimental method relies on a change of equipment, this requires a substantial financial investment. Equipment is not only a financial cost but a time cost as well: the proper use of the equipment likely involves a great deal of tacit knowledge, which can take a long time for a lab to develop. As a result, a lab is likely to try and maximize the value of their existing equipment unless there are very large gains from a new piece of equipment.

A third cost is that of collaboration. Simply working with people in a different physical location is costly—communication is more difficult, and this can make coordination more challenging. As Galison (1997) discusses, this can be particularly true when working across disciplinary (or sub-disciplinary) boundaries. There may be different standards of rigor, different definitions of terms, and different sets of standard approaches across disciplines, and these differences add to the communication cost. Furthermore, when there are multiple PIs involved in research, the credit for discovery has to be shared. The more labs are involved, the less credit any given PI can take for the conducted research. It also stands to reason that for the vast majority of cases, collaboration is a substantially lower cost than skills investment.

It is worth noting that the first two costs, skills acquisition and equipment, are *fixed costs*. The lab pays for them regardless of their relationship to output. Even if a lab never uses a machine, it still pays for it. Collaboration, on the other hand, is a *variable cost*. Labs incur the cost in relation to outputs. Collaborations are, by their nature, project oriented. The costs come from the production of the collaboration. Whereas skill acquisition has an upfront fixed cost, and then can be costlessly deployed on as many projects as one wishes.[10]

The nature of these costs suggests that the optimal strategy for a rational agent looking to maximize her return on investment is to minimize the marginal cost

[10] The reader may notice that I allow the cost of actually engaging in research to drop out of my discussion. I do this for two reasons. One, it is difficult to in the abstract account for the cost of a particular research program. Two, nothing I propose should have a significant effect on day-to-day effort expended by a scientist that isn't captured by the costs I have identified. While day-to-day costs of research are surely not zero, it's not clear that they have a functional relationship at the abstract level with the costs I do consider. Some equipment may make certain tasks easier or less time consuming, but I am thinking of equipment here as the capital version of skills—enabling technologies rather than efficiency-enhancing technologies. These costs are washed away to simplify the model without changing the basic dynamics.

of research. What does this imply? That in an ideal world, a scientist would get as much out of her fixed-cost assets as possible without turning to anything with variable costs. So, in the ideal world, a scientist facing this cost structure would avoid collaboration with others, and only engage in those projects for which her existing skills and equipment are suited. In a slightly less ideal world, the space of projects is another limited resource: low hanging fruit that only requires the particular skills and equipment that a scientist currently possesses will be plucked over time. As this happens, scientists will be driven to collaborate with others or accept the large fixed costs of further training or equipment purchase. Depending on the relative costs, we will either see more skills investment on the part of scientists or greater levels of collaboration. If scientists have different individual costs for each, we may see some inter-agent variation.[11]

This is an inversion of standard assumptions. The Weisberg and Muldoon model suggests that scientists have research problems in mind and then search around for the tools necessary to investigate them. Kitcher, Strevens, and Zollman offer no particular differentiation of the skills of scientists, only differentiation of which projects or hypotheses they choose. Here, we find instead that scientists start with the tools and then look for the projects that suit their tools. This is a large and important shift. On this approach, scientists opportunistically seek out projects for which their skills are suited.

This shift has several consequences. First, it continues the trend in the literature of moving away from the view that scientists operate as ideal individual epistemic agents. Not only do scientists care about non-epistemic concerns, such as fame or esteem, but they are also a bit more mercenary about achieving their goals. On this account, scientists can be thought of as optimizing the epistemic return on their lab's investments. While this may initially appear to be distasteful—scientists reduced to economic agents going after epistemic returns in the same way firms go after profits—it does help us make sense of the empirical patterns that we have seen in academic science.[12] Not only does it help make descriptive sense, it also offers normative guidance: scientists *should* exploit their existing skill sets rather than seek out new ones. Our scientific output is greatly increased when this strategy is taken on.

[11] This can be a distinction between persons or disciplines. In the personal interpretation, some individuals may be quick studies and are able to learn new techniques fairly easily, while others may not be so fortunate. Likewise, some people have superior networking skills and are better able to generate productive collaborations than others. In the disciplinary interpretation, we may find that learning new skills or techniques is easier or harder. It may be that in math, which has a low number of coauthors per paper, learning a new technique is relatively inexpensive, compared with, say, experimental biology, which has a higher average number of coauthors per paper. It's also possible that biologists have better people skills than mathematicians, which makes larger collaborations easier.

[12] We can also lessen how mercenary this appears by thinking of the epistemic returns as the scientist's utility, which could be over contributions to interesting projects, or fame, or whatever else the scientist might value. One might suppose that the choice of utility function is at least a little constrained by the demands of tenure and successful grant-writing, but this doesn't prohibit a scientist from favoring her own interests when considering projects that suit her skills.

In a world of plentiful scientists and hard problems, it is remarkably wasteful to have each scientist be a generalist—the training period would be enormous, and we would have alarming amounts of duplicated effort. As wonderful as scientific competition is for instigating scientists to work harder and faster, too much competition is socially wasteful. This is part of Kitcher's original point: if everyone worked on the most promising project, there would be significant and socially wasteful overlap. The marginal scientist would not be adding much to the output of science. But his insight is deeper still: we want to see scientists not only work on different projects but also acquire different skills to bring to those projects. If everyone shared the same skill set, they would be limited to the same set of problems.[13]

If we instead allow scientists to specialize and exploit their particular skill sets, then we avoid this problem. Scientists can explore the problems that are suited to their skills and ignore those problems that they cannot productively engage with. On this account, there is an initial investment choice for a scientist to pay some fixed costs (during her training period and initial lab start-up), and then we enter a period of skill exploitation. Our scientist will attempt to exhaust the set of problems for which she has a comparative advantage in solving. If this set of problems were to run out, or become sufficiently low value, she could then either choose to cooperate or attempt to seek a different set of problems.

4.5 Problem Space as a Determinant of the Social Structure of Science

This approach to thinking about the social structure of science is continuing in a trajectory established first by Kitcher, and then furthered by Weisberg and Muldoon: the structure of problems matter to understanding how scientists ought to behave. Kitcher introduced the idea of a success function for a problem, which allowed us to think about how many scientists could productively work on a single task. Weisberg and Muldoon introduced a landscape metaphor for scientific discovery, arguing that scientists had to discover the most fruitful areas of science to explore. The approach here goes further in arguing that scientist behavior is a function of the problem space that scientists face.

[13] Here again we see the insight about increasing the epistemic productivity of the marginal scientist intersecting with Adam Smith's point about the division of labor being dependent on the extent of the market. If we think about few problems and few scientists, as has been common in the modeling literature, it is easy to miss the importance of the division of labor in the case of joint production, precisely because it only becomes a powerful tool once there is a larger population of scientists working on a large potential set of problems. The wider set of problems gives scientists a reason to invest in a narrower set of skills, confident that there will be enough problems that one can apply that set of skills to, whereas the large population of scientists is compatible both with individual specialization and community-level coverage of the full range of available problems. Too few problems, and we don't get the reason to specialize. Too few scientists, and we lose good coverage of scientific problems.

The problem space is simply the set of available problems for scientists. The full set may or may not be known, nor is it necessarily the case that it is transparent to scientists what skills are required for a given problem. But I do suppose that scientists, like any skilled people, have a sense of what problems will be amenable to their skill set. This may be imperfect, of course, but good enough to engage in a search of problem space. This search process is often built into research labs: journal clubs usually involve looking through recently published research and discussing it. Part of the function of a journal club is to help train graduate students, but part of it is to look for new problems that can be worked on. By reading journal articles, research labs can get a sense of whether their skills would be appropriate for the projects that might surround the published result. It also exposes them to possible collaborators. Likewise, conferences and workshops perform a similar role—they are a relatively efficient way for a lab to be exposed to the current state of the art, paired with a mechanism for meeting existing and potential collaborators. These mechanisms, and others like them, allow scientists to remain at least familiar with the available problem space with relatively low overhead costs. So, even if scientists do not have perfect epistemic access to the full problem set that may be awaiting them, they still have the means to select between available problems, making their selections based on which they think they are equipped to investigate, and how interesting they think the problem is.

The problems that scientists are exposed to give them a domain of application for their skill set. Given this domain, scientists can determine a set of problems to work on. The hardness of problems, as defined by the sets of skills required for attacking the problem, determines the level of cooperation amongst scientists. The problem space is also a way of describing a particular discipline. The problems that scientists face in, say, high-energy physics, give rise to a set of tools that scientists invest in to help investigate them. The problems in population ecology are of a different nature and so demand a different set of skills.

This approach, in which disciplines are determined by a set of problems, and the skills associated with the discipline come out of the skills found necessary to work on those problems, helps us account for the different levels of cooperation found between different disciplines, and also why we find some disciplines colonizing others. We can also see how, over time, what skills considered "core" to a discipline can evolve as the set of active problems change.

4.6 Cooperation and Colonization

With a clearer handle on the costs faced by scientists, and the shape of the problem space they face, we can think about two significant phenomena in science: cooperation between scientists, and the colonization of fields like biology by fields like physics. In each case, we can think of these phenomena in terms of rational responses that scientists take to the current problem space.

In the case of cooperation, the picture is straightforward: as low-hanging fruit runs out, scientists can only work on papers if they collaborate with others or invest in more skill acquisition. Typically the cost of skill acquisition will be too high, and so collaboration is the rational response.[14] The rate of collaboration and number of collaborators will be driven by the problems that scientists can identify. We can suppose that scientists will only take on as many collaborators as are needed and will look to minimize the total number of collaborators on a project. After all, the more collaborators, the more dispersed the credit for a publication. We can also reasonably suppose that as the number of collaborators increases, the more difficult it is to successfully collaborate: there are more veto points in decision-making, it is harder to write a joint paper, and coordination in general becomes more challenging as group size increases.

To add a bit of nuance to the discussion, we can also move away from averages. In particular, we might see different patterns of collaboration within a given discipline depending on the distribution of particular skills amongst scientists and the requirements of the problem space. It may be that some specializations in a discipline have plenty of problems amenable to them, while others need collaborators. Further, some skills, like data analysis, may be solely collaborative.

Collaboration, then, is a straightforward response to a problem space that is of higher average difficulty, where specialization is required for reasonable training times. Scientists can only be skilled at so many things, and the problems get harder over time as easier ones are solved. Individual scientists might find the need to collaborate sooner or later than their peers depending on the particular skills they have and the nature of the problem space they face. But what happens if collaboration becomes overly burdensome? It may be the case that a scientist's skills are undervalued in her current field, because of a glut of such skills, which may mean she is unable to reliably secure worthwhile collaborations. Or it may be that the size of collaborations has grown too unwieldy. What is such a scientist to do?

One option is to attempt to exit her field and enter another. In particular, she ought to enter a field in which her skill set is rarer but is readily applicable to the problem space. In this way, a scientist can immediately increase the value of her existing investment. Rather than toil away at low-value work, she can work on projects of higher value, just in a different domain.

As a simple (somewhat stylized) example of such a phenomenon, consider physics. While physics is certainly a very active field with many interesting problems to work on, many of those problems are increasingly requiring large-scale equipment, such as the Large Hadron Collider or the LIGO project.[15] These sorts of projects frequently produce publications with hundreds of coauthors. This is a challenge for a junior person to build up a reputation on her own—how does she identify her contribution to such a massively collaborative project? Increasingly, physicists have begun to work on problems that are more traditionally associated with other fields, such as biology and social science.

[14] The special case here is, of course, when new methods or tools need to be developed, such that narrower, deeper specialization becomes necessary. When no one yet has a needed skill or piece of equipment, no amount of collaboration solves the problem. More investment is needed first.

[15] The Large Hadron Collider is a very large particle collider for high-energy physics research, built with the aim of finding the Higgs boson. The LIGO (Laser Interferometer Gravitational wave Observatory) project was likewise built with the aim of detecting gravitational waves. Both projects require very large, complex machines and very large teams of scientists.

Typical training in physics gives physicists excellent modeling skills, very strong quantitative skills in general, and exposure to a number of data analysis techniques. These are reasonably domain-neutral skills and so are readily applicable to other fields. In neuroscience, for instance, physicists have applied coupled oscillator models to describe neuronal firing patterns. Understanding of coupled oscillators is simple for physics, but the application to neuroscience is novel. Many other such examples exist. In such a way, scientists can exploit their existing skill set by seeking problems in different areas.

A potential challenge to this colonization approach would be that it might entail large switching costs: a physicist moving into biology might need to learn enough biology to publish in biology journals, for example. This may be equally (or more) costly as picking up a new skill set in physics. While no doubt this can be the case sometimes, it is equally plausible for physicists to simply publish in physics journals, for an audience of physicists. Or, as we find happening now, creating new academic departments like biophysics. These moves minimize switching costs, as physicists can continue to write for an audience of physicists about the kinds of problems physicists are interested in. There may be higher impact opportunities for engaging more directly with biologists, and so we may expect to see some investment in that direction as well.

This same story can point to why we see significantly less movement from biology to physics—while a number of physics skills are applicable to parts of biology, there are fewer tools in biology that apply to physics. General evolutionary dynamics can be found in many fields, which may lead a theoretical biologist to branch out beyond biology, but a typical experimental biologist is probably not going to be able to see much value in attempting to switch to high-energy physics.

Disciplinary colonization is thus a continuation of the story of maximizing the return on a scientist's skills investment. By employing a simple decision framework to take into account the serious costs involved in project selection, we can more clearly capture not just the division of cognitive labor within a discipline, but we can also begin to see how movement between disciplines occurs. Scientists will flow from problem spaces with low rewards for their skill sets to problem spaces with higher rewards for their skill sets.

4.7 Conclusion

I have argued that previous models of the social structure of science have neglected crucial features of real scientific practice. In particular, they could not account for scientific collaboration, and as such, they could not have a proper account of the division of cognitive labor in science. I have offered a framework that can generate these features of science by taking into account the real costs that scientists face in their work. In doing so, this framework also offers an account of some larger features of science, such as disciplinary colonization.

While improving the descriptive state of our understanding of the social structure of science is important, a reader may ask whether this is sufficient. After all, the models of the social structure of science I have discussed in this chapter are not primarily meant to be descriptive accounts of science, but instead stylized models of science that serve to make normative claims about the social structure of science. What is the normative contribution of the framework that I have suggested?

I see the primary normative contribution as falling out of the improved understanding of the division of cognitive labor. Weisberg and Muldoon pointed out that it is not just what evidence scientists have, but how they respond to it that can determine how successful a community of scientists can be. In this framework, we see that skill diversity in science is equally important. As problems become more difficult, individual labs can't be enough. Collaboration with labs that have the same set of skills is not very useful. Instead, we need a wide variety of skills available in scientific communities.

This has a few important consequences. First, it suggests that tolerance of methodological pluralism is a potentially important virtue for scientific communities. Unless those communities can predict the nature of the skills necessary to solve problems, it is in the communities' own interest to promote skill diversity. This benefits everyone in a collaborative setting, as it helps to ensure the resources needed to tackle increasingly complex problems. Second, we might encourage, rather than lament, ever-increasing levels of specialization. If Smith is right, we should expect more specialization, paired with more collaboration, to result in increased scientific productivity. There may be a natural limit in some disciplines for how finely one can divide labor, but it is far from clear that we are approaching that limit. Third, and perhaps most importantly, this suggests that we need more investment in and understanding of collaborative science. Skill diversity becomes a necessity given the nature of the increasing complexity of scientific discovery. But with specialization, we can run into challenges of differing standards of rigor, different jargon, and the requirement for scientists to have the acumen to navigate these challenges, in addition to being skilled at the science itself. Galison has begun this discussion with his study of trading zones, but there is more work to do. The division of cognitive labor means divided interests, standards, and divided understanding. An essential project is in understanding how the division of cognitive labor can still result in rigorous production of knowledge by the community of science. By examining the constraints of scientists, we can better see both how some of the social structure of science emerges and how we can improve scientific production.

References

De Langhe, R. (2014). "A unified model of the division of cognitive labor." *Philosophy of Science* 81 (3), 444–459.

Galison, P. (1997). *Image & logic: A material culture of microphysics*. Chicago: University of Chicago Press.

Glanzel, W. and A. Schubert (2005). "Analyzing scientific networks through co-authorship." In H. F. Moed et al. (eds.), *Handbook of quantitative science and technology research*, 257–276. Amsterdam: Springer.

Kitcher, P. (1990). "The division of cognitive labor." *Philosophy of Science* 87, 5–22.

———. (1993). *The advancement of science*. New York: Oxford University Press.

Muldoon. (2013). "Diversity and the division of cognitive labor." *Philosophy Compass* 8 (2), 117–125.

National Science Foundation. (2012). *Science and engineering indicators 2012*. Available at http://www.nsf.gov/statistics/seind12/.

Rockey, S. (2012). "Our commitment to supporting the next generation," in NIH Extramural Nexus, Office of Extramural Research. Available at http://nexus.od.nih.gov/all/2012/02/03/our-commitment-to-supporting-the-next-generation/.
Smith, A. (2003). *An enquiry into the nature and causes of the wealth of nations*. New York: Bantam Classics.
Strevens, M. (2003). "The role of the priority rule in science." *Journal of Philosophy* 100 (2), 55–79.
Weisberg, M. and R. Muldoon. (2009). "Epistemic landscapes and the division of cognitive labor." *Philosophy of Science* 76 (2), 225–252.
Wray, K. B. (2014). *Kuhn's evolutionary social epistemology*. New York: Cambridge University Press.
Zollman, K. J. S. (2010). "The epistemic benefit of transient diversity." *Erkenntnis* 72, 17–35.

PART THREE

AUTHORSHIP AND REFEREEING IN COLLABORATIVE RESEARCH

5

Making an Author in Radically Collaborative Research

BRYCE HUEBNER, REBECCA KUKLA, AND ERIC WINSBERG

Collaborative authorship in science is the overwhelming norm, and articles with tens or even hundreds of listed authors are not uncommon. Across philosophy, there has been an increasing recognition that collaborative research and collaborative knowledge production are central epistemological practices and not just fringe phenomena. The philosophical literature on collective actions and intentions, collaborative knowledge, and group minds is exploding (Gilbert 1989, 1996; Huebner 2013; List & Pettit 2011; Rolin 2008; Thagard 2002; Tuomela 2002; Wray 2002, 2006). In spite of this focus on collective and collaborative phenomena, we think that questions about collaborative *authorship* raise distinct philosophical issues that have yet to received sufficient attention; sometimes these have been improperly assimilated to issues concerning collaborative knowledge and action. In this chapter, we examine the unique difficulties inherent in establishing authorship in the context of collaborative research.

Authoring is not the same as knowing, though it cannot be understood independent of its relationship to the epistemic activities responsible for producing knowledge. Although *knowing* is not an inherently social or communicative notion, authorship is by nature social and communicative, even where it is not collaborative. Unlike mere knowers, (nonfiction) authors are *accountable* for the content and accuracy of the information they produce. An author must be able to justify and vouch for the truth of her claims, defend her products when challenged, and retract her claims when she cannot defend them in light of criticisms or new information. An author might make false claims, at least so long as she remains accountable and takes proper responsibility for them. Inquiries backed up by faulty processes of knowledge production and acquisition yield authors that are less reliable sources of information. But such failings are not failures in the status of *authorship*. Failing to know is not the same as failing to author.

Collaborative research can be organized in a variety of different ways, and it can be performed in different material contexts. Not everyone involved in a research project is an author; indeed, we argue that people are sometimes listed as 'authors' on a publication when they should not be properly understood as authors at all.

Boyer-Kassem, Mayo-Wilson, & Weisberg. *Scientific Collaboration and Collective Knowledge*.

We examine three broad classes of research architectures using case studies. We consider collaborative research with multiple authors; collaborative research with a single author and many collaborators; and "radically collaborative" research that is distributed widely over time and space, involves researchers from multiple disciplines, and typically results in publications with dozens of listed authors (cf. Kukla 2012; Winsberg 2012; Winsberg, Huebner, & Kukla 2014). We hold that the first two types of collaborative research leave our basic understanding of authorship untouched. But we argue that an important range of research questions require radical collaborations that seriously challenge our understanding of and standards for authorship.

There are currently no good strategies for understanding or ensuring accountable authorship in the context of radical collaborative research. If the results reported by such collaborations are to be authored at all, they must be *group authored*. But we contend that a great deal of radically collaborative research has structural features that make group authorship impossible, thus making authorship *simpliciter* impossible in these cases. Even where there are good-faith attempts to produce group authorship, major architectural barriers remain. This, we claim, poses concerns about epistemic security as well as our ability to interpret and decide when to trust the results that are reported in radically collaborative publications.

5.1 Catch-and-Toss Authorship and Centralized Control Authorship

We begin by looking briefly at two kinds of collaborative research that do not pose a fundamental challenge to a traditional notion of authorship. First, we consider a type of collaborative research where a few authors work together to produce a single coauthored paper—much as we have done with this one. Borrowing a term from Andy Clark (1998, 105–6), we call this "catch-and-toss" authorship. Second, we examine a kind of collective research that relies on a widely distributed form of information processing, but which retains a high degree of centralized control over the reported results. We use the first type of case to demonstrate a way in which research can be deeply social without compromising authorship; we use the second type of case to demonstrate a critical respect in which the distribution of information processing does not entail the distribution of authorship. With these cases in hand, we turn in the following section to a discussion of the kinds of radical collaboration that put more significant pressures on our ordinary understanding of authorship.

Let's begin with the kind of case most familiar to philosophers. When philosophers collaborate, they typically exchange ideas in a way that allows each author to retain epistemic authority over the resulting paper. In the most familiar form of philosophical collaboration, one author develops an idea (either in discussion or as she writes her section of a paper). She then "tosses" it to another author who "catches" it, revises or extends it, and then tosses it back. This type of collaborative writing often consists in a long series of such catchings and tossings, and the process is typically repeated until every member of the collaboration is happy to put her name on the resulting paper. These types of collaborations allow every author to be accountable for her own contributions. But, more importantly, since the writing and thinking are genuinely *interactive*, each author is likely to form a relatively clear sense of the structure of the overall argument. While there may remain some minor disagreements regarding what ought to be said, each author remains in position to vouch for the finished product. Each author is typically in a position to defend the claims produced through the collaboration when they are challenged, and each author is in a position to retract the claims that cannot be defended in light of criticism.

Of course, things are not generally this simple, and there is a different form of catch-and-toss process that arises when a project requires a unique technical skill or information that is only available to a specialist. In some cases complex statistical analyses are required that a lead author does not know how to carry out; in other cases, the relevant research may require a working knowledge of some range of scientific data that is only available to a person who has undergone special training. But, in these cases, a single author can do the work that is relevant to a particular section of a paper in a way that allows her to retain responsibility for her portion of the paper. For example, one author may catch a data set collected by someone else, carry out a complicated statistical analyses, and toss these results back to the lead author who inserts them into the paper.[1] Clear lines of responsibility are maintained in this type of coauthorship, and where the corresponding author is unable to vouch for a claim that is made in the paper she will know which collaborator has this capacity. This means that it is possible for collaborations to include specialists while still making sure that there is someone who is epistemically accountable.

In both of these cases, catch-and-toss collaborations help organize information in a way that functionally distributes the labor required for producing some purported piece of knowledge. The ideas produced through catch-and-toss collaborations are often only possible because of the patterns of reciprocal feedback that arise between collaborators. However, each part of the coauthored paper is the product of an individual author who produces representations that can be tossed back and forth.

[1] Although catch-and-toss collaborations fall reasonably comfortably within our traditional understandings of authorship, they can still create epistemic complications. When people come to a project with complementary forms of knowledge and expertise, which are drawn from different fields of inquiry, they must somehow find a way to negotiate the details of a collaboratively authored project. Precisely how this occurs differs from collaboration to collaboration. At one extreme, there are collaborations in which multiple revisions of a project are carried out, with an emerging paper being tossed back and forth until every collaborator understands and agrees to every one of the claims that are made. This yields a situation where every author fully understands the target phenomena and can thereby be seen as an author in the traditional sense of the term. At the other extreme, there are collaborations in which every author simply accepts the claims that are made by collaborators, developing no further understanding of the target phenomena beyond their original field of expertise. This yields a situation where authorship is distributed across the members of a group, but where it is possible to recover structures of accountability by tracking the patterns of trust and acceptance within the group. As Hanne Andersen & Susann Wagenknecht (2013, 1892) note, collaborative scientific practice typically lies somewhere between these extremes. Even in small-scale collaborations "scientists in some areas learn from each other and in others remain epistemically dependent." So, the social epistemology of even small-scale catch-and-toss collaborations can be more vexed than we have suggested here.

This yields a sort of joint authorship in which claims to knowledge are produced through a highly interactive process that leaves standard structures of accountability and responsibility in place.

But there are types of research, and accompanying forms of collaboration, that are too complex for this catch-and-toss model to be practically possible. For instance, after the French Revolution, the Académie des Sciences famously instituted a metric system of weights and measures. They also stipulated that right angles would now be divided into 100 grades rather than 90 degrees.[2] Nautical navigation relied heavily on trigonometric tables, and using this new standard of measurement on the fly proved practically impossible. So, the Académie instituted the Bureau du Cadastre to construct new trigonometric tables using grades rather than degrees. The director of the Bureau, Gaspard de Prony, was enamored of Adam Smith's idea of "distributing labor," and he hired a team of human computers to carry out the arduous task of constructing the trigonometric tables. The vast majority of these computers (around 90 of them) "were former servants and hairdressers, who had lost their jobs when the Revolution rendered the elegant styles of Louis XVI unfashionable or even treasonous" (Grier 2005, 36). As such, they had no mathematical training beyond the basic abilities required to add and subtract. So, a team of eight "planners" supplied these computers with worksheets that allowed them to carry out simple calculations; the planners then took differences between the calculated values to spot-check for errors. Finally, a small number of trained scientists figured out the appropriate formulas to be passed down to the planners. All the relevant calculations were eventually passed along to de Prony, who put together a 19-volume manuscript that included all the tables.[3]

Structurally speaking, this kind of research shares much with the catch-and-toss collaborations discussed above. However, in this case, the human computers became nothing more than sources of information for de Prony's mathematical investigations. The organization and distribution of intellectual labor instituted clear lines of responsibility, such that de Prony (and perhaps his team of trained scientists) retained accountability for the resulting tables. The human computers became mere tools, to be used in the service of a project carried out by a few highly trained and highly intelligent scientists. Because of the organization of this research, de Prony (and perhaps the trained scientists) remained epistemically accountable. These authors could vouch for the results of this massive collaboration; they were the ones who were epistemically accountable for producing accurate tables, defending them if challenged, and revising them if necessary. The relevant structures of epistemic responsibility and accountability were maintained in the Bureau du Cadastre to allow for the distribution of collective research while leaving traditional notions of authorship intact. In effect, the text was still single-authored, even in the face of distributed information processing, because one person retained centralized control over the research process, including its methodological standards and

[2] Our discussion of De Prony in this paragraph follows Grier (2005, 34–38). For an interesting discussion of this research in another context, see Shieber (forthcoming).

[3] Unfortunately, the manuscript was never published because the publisher went bankrupt, and the Napoleonic government had no interest in publishing the volume.

implementation. While many people participated in the production of knowledge, only one person had the status of the author of the document communicating that knowledge.[4]

5.2 Genuinely Distributed Epistemic Labor

In the type of case we just examined, labor is distributed, but there is no need for lower-level agents to exercise special epistemic skills or judgment, because the author who retains centralized control can provide lower-level agents with detailed and specific instructions about how to handle every problem that they might encounter. We now turn to a set of cases involving distributed epistemic labor, in which the primary reason for involving a large number of actors is that no single actor *can* possess the relevant knowledge and skills necessary to *produce and sanction* the desired knowledge claims. In such cases, multiple actors must exercise special epistemic skills and judgment. In the types of collaborations that interest us here, there is no one who understands the role that is played by every researcher, and there is no one who knows what everyone else has contributed to the project; there is no one who has even testimonial knowledge that the other researchers are trustworthy; there is no one who has command over how the various pieces of the study fit into a coherent whole.

We argue that the reports that result from these types of radical collaborations cannot be understood as multiply authored, as in the catch-and-toss model, or single-authored while involving distributed labor, as in the centralized control model. Instead, if they have any kind of author at all, it must be a group author. Unfortunately, as we will see, the architecture of radically collaborative research is not typically of the right sort to produce a group author. Nor is it at all clear, given the real pressures and constraints that such research faces, how it could be restructured to enable group authorship.

5.2.1 Modeling in Climate Science

No case exemplifies radically distributed epistemic agency better than climate science—especially when it involves massive, modular, and highly complex *coupled atmosphere-ocean global climate model* (AOGCM) simulations, such as the National Oceanic and Atmospheric Administration's GFDL CM2.x Model.[5] CM2.x contains over a million lines of code, with over a thousand different parameter options. Its novel component modules were written by more than 100 people. These modules are constantly changing, and they involve hundreds of initialization files that contain incomplete documentation. Loading input data into a simulation run takes over two hours. It takes weeks to produce a model run out to the year 2100, using more than 100 processors

[4] Strikingly, a similar model of highly distributed, centrally controlled research was instituted at the Oak Ridge National Laboratory as a way of isolating isotopes of Uranium-235. For a detailed discussion of this case see Winsberg, Huebner, & Kukla (2014).

[5] Details concerning the CM2.x model come from Dunne (2006).

running in parallel; and it takes months for them to reproduce thousands of years of paleoclimate data. If you store the data from a state-of-the-art GCM every five minutes, it produces tens of terabytes of data per model year. But these data must be collected relatively quickly to be of any use. Policymakers want immediate answers to near-term possible climate changes, and we would like to know what characteristics the climate is likely to have in 2050 sometime before 2050. This is part of the reason why these climate models must reflect the work of hundreds of researchers working at different times and in different physical locations. Furthermore, the time-indexed nature of the research makes it inherently unreproducible.

The sheer size and complexity of such models make it clear why their construction and sanctioning must involve a vast army of specialists. Their production and use requires expertise in climatology, meteorology, atmospheric dynamics, atmospheric physics, atmospheric chemistry, solar physics, historical climatology, geophysics, geochemistry, geology, soil science, oceanography, glaciology, paleoclimatology, ecology, biogeography, biochemistry, computer science, mathematical and numerical modeling, statistics, time-series analysis, and more. Furthermore, no GCM is built from the ground up in one short surveyable unit of time. They all rely on assemblages of methods, modules, parameterization schemes, initial data packages, bits of code, coupling schemes, and so forth that have been built, tested, evaluated, and credentialed over years or even decades of work by climate scientists, mathematicians, and computer scientists of all stripes.[6] This yields a sort of "fuzzy modularity" in these models (Lenhard & Winsberg 2010). In current atmospheric GCMs, we find a deliberately modular architecture that consists of coupled models, which are themselves comprised of numerous interactive submodels (e.g., modules for general circulation of the atmosphere, cloud formation, dynamics of sea and land ice, and effects of vegetation). The interactions between these models determine the global dynamics in simulation runs, and interactions can be quite complex.[7] In a modular system, individual models (ideally!) act as interchangeable but highly networked modules. The results of such modules can first be gathered independently and then synthesized. However, in current GCMs, the data is continuously exchanged between modules during the runtime of the simulation.[8] Thus the dynamics of the model can only be understood as the complex result of the interaction between these modules, each of which is infused with the judgments made by numerous different scientists, on the basis of their own assumptions about what the right way to parameterize is. To say

[6] There has been a move, in recent years, to eliminate "legacy code" from climate models. Even though this may have been achieved in some models (this claim is sometimes made about CM2), it is worth noting that there is a large difference between coding a model from scratch and building it from scratch, that is, devising and sanctioning from scratch all the elements of a model.

[7] To take one example, the coupling of atmospheric and oceanic circulation models is recognized as a milestone of climate modeling. Both components had an independent modeling history, including independent calibrations of their respective performance. Putting them together was a difficult task because the two submodels interfered dynamically with one another. For an account of the controversies surrounding early coupling, see Shackley et al. (1999); for a brief history of modeling advances, see Weart (2010).

[8] Because data are being continuously exchanged one can accurately describe the models as parallel rather than serial in the sense discussed in Winsberg (2006).

that such models have only a sort of fuzzy modularity is to note that the interactivity and the balancing of approximations prevents the system from being broken down into separately manageable pieces.

The problems for authorship that are evoked by this type of fuzzy modularity are exacerbated by the fact that epistemically salient decisions in climate modeling must be sensitive to a continuous, dynamic flow of information through the model. The collaborators must carry out continuous deliberative adjustments in light of new circumstances and new information, and this must frequently occur without the time or means to consult with other members of the research team. This is why the operation of each module must rely on a mixture of principled science and decisions about parameterization. Climate modeling involves literally thousands of methodological choices. Many crucial processes are poorly understood; many compromises must be made in the name of computational exigency; and so forth.

But no methodological choice can be defended in a value vacuum. When one asks, "Why parameterize this process rather than try to resolve it on the grid?" or "Why use this method for modeling cloud formation?" the answer is rarely "because that choice is simply objectively better than the alternatives." Most of these choices are better in some respects and worse in others, and the preference for the one choice over its alternatives always reflects the judgment that this or that respect is more important. Some choices will result in more accurate predictions with respect to one variable (e.g., tropical precipitation), while its alternative will be more accurate with respect to another (e.g., polar ice). Some choices will increase the probability of finding a certain degree of climate variation; other will do the opposite. So any rational reconstruction of the history of climate science (if it existed) would have to mention each of these predictive preferences at pain of making most choices seem completely arbitrary (Biddle & Winsberg 2009; Lenhard & Winsberg 2010; Winsberg 2010).

Meanwhile, climate experts, in light of the individually limited role that they play in the socially extended activity of building climate knowledge, can only arrive at opinions about the future of the climate in ways that are fundamentally mediated by the complex models that they build. And they are incapable of sorting out the ways in which past methodological choices made by other scientists—scientists whose expertise they don't entirely share—are influencing, through their entrenchment in the very models that mediate their inferences, the conclusions they deliver to policymakers.

No single person is in a position to offer a rational reconstruction of a climate model. Too many of the methodological choices are buried in the historical past under the complexity, distribution, and historically path-dependent character of climate models. The various local standards and values employed in climate science lie in the nooks and crannies (Winsberg 2012). They might very well have been opaque to the actors who put them there, and they are certainly opaque to those who stand at the end of the long, distributed, and path-dependent process of model construction. There is no one who could, even in principle, be held accountable for the claims to knowledge that are produced using CM2.x. To put the point another way, there is no person who has the requisite expertise to play the role of de Prony in these modeling activities, and accordingly there is no single person who can legitimately be treated as the author of these claims.[9]

5.2.2 Multisite Clinical Trials in Biomedical Research

Multisite clinical trials, the purported gold standard of biomedical research, create an analogous pattern of difficulties. In many cases, biomedical research takes the form of geographically distributed projects that involve many teams of researchers, each of which uses nurses, lab techs, students, departmental administrators, and many others to help implement the study and keep track of data. Papers with dozens of authors, often from several continents, are the norm. In many cases, no one has the expertise and training to design and implement every part of the study. Such research often draws on a wide variety of disciplinary expertise, including biostatistics, cell biology, oncology, and immunology, among many others.[10] These research projects are so dispersed and so multidisciplinary that no one can have access to more than a small corner of these data. Furthermore, no one can have expertise in all the on-the-ground skills required to administer tests, read results, recruit subjects, and so forth. It is unlikely that the biostatisticians and immunologists fully understand each other's contributions, or that researchers in France know exactly how subjects were recruited in Taiwan, and so on. This is decentralized, distributed collaborative research on a massive scale—a scale frankly unthinkable until recent developments in communication technology.

[9] Contrast this with even a very complex example of catch-and-toss authorship: Timothy Gowers and Michael Nielsen (2009) report the results of a recent case collaborative research in mathematics known as the Polymath Project. The aim of the project was to find an elementary proof of the Hales-Jewett theorem. Over the course of 37 days, 27 mathematicians contributed approximately 800 substantive comments to a collaborative discussion, yielding 170,000 words of text, and an elementary proof of the Hales-Jewett theorem. Contributions to the proof were made by world-renowned mathematicians and university professors, as well as high school math teachers. Furthermore, contributors were active in the discussion to very different degrees, and at different points over the course of the collaboration. This case provides a nice set of insights into the operation of catch-and-toss collaborations. The discussion reveals that ideas are often proposed, then adjusted, and sometimes discarded over the course of the collaboration. But, perhaps more importantly, it reveals that advances in understanding often result through the "aggregation and refinement of many smaller insights" (Gowers & Nielsen 2009, 880). This process generates practical questions about authorship. For example, how should credit be assigned when a contributor offers a single insightful contribution, and how should it be assigned when a contributor is prolific but not particularly insightful? But a working record of the collaboration remains available for outside scrutiny. If there were a mistake in the proof, it could be tracked down, and it could be determined how mistakes were introduced, and how they were perpetuated in the project. This may not be an easy task, but at least it would be possible.

[10] The December 2011 issue of the *New England Journal of Medicine* includes five "original research" articles, with 27, 6, 29, 19, and 13 authors respectively. All but one are written on behalf of a much larger research group, and links to a supplementary appendix listing hundreds more collaborators and participating investigators. All five articles list authors from various countries and various institutions within each country, and each contains authors with diverse disciplinary backgrounds. Other issues display roughly similar authorship patterns.

The problem is not merely that there is no person or small group of people that could have the time or the expertise to run the study on their own. The data in biomedical research, as in climate science, are often time and place dependent, and questions about clinical effects are often context specific. Policymakers frequently need to know immediately whether to close schools in response to potential epidemics, and there are sometimes political pressures to approve a drug or remove it from the market as soon as possible. Because medical interventions often work differently in different populations living under different social and material conditions, it is common for a research question in biomedicine to *require* trials in far-flung locations, which must be conducted under messy and chaotic conditions, in the face of communication challenges. Additionally, since bodies change quickly, and since harms and benefits from a treatment often evolve over time, the time frame of a study (how quickly it must be completed, how close together in time the collection of different data sets must be) is often fixed by the research question. In other words the multisite distribution of studies along with their rapid time frame is often essential to the research design. This, again, makes these studies inherently unreproducible.

Unlike centralized control cases, researchers involved in this sort of radically collaborative, distributed research must often make methodological adjustments on the fly in response to noncompliant research participants, unforeseen barriers to implementation and communication, surprising side effects, and so forth. It is hard to know in advance what methodological uncertainties, judgment calls and choice points collaborators will face. These adjustments may be made differently at different sites and in different ways at different stages of the research process. Whereas de Prony's computers repeated the same task under controlled conditions, these researchers must draw on their own expertise to make on-the-spot decisions about how to proceed under imperfect and sometimes surprising conditions. For example, when researchers encounter participants who are partially compliant, who disappear halfway through a study, who display ambiguous symptoms, who are difficult to communicate with, and so on, they must still decide whether to use data from these participants. But this requires judgment calls about how much to bend protocol to get through the study.

For all these reasons, there is often no one involved in the research who has more than a loose idea who is doing what parts of the sprawling project, using what methods, or how accurately. Whatever being an "author" comes to in the context of climate science or multisite clinical trials, it shares little with the traditional understanding of scientific authorship. Indeed, actual practices of assigning authorship make no pretense that authorship tracks responsibility for the production or the justification of the results. There is no presumption that the actual writing of the article, which many scientists see as extra-scientific busywork, will be performed or even overseen by the lead "authors."[11] Authorship constitutes an institutional status; it does not represent a specific form of epistemic labor. Authorship is assigned and ordered according to whose grant money was used, who runs the lab, who has tenure, who needs a job, and so forth.

Often this reality is stubbornly denied by the researchers themselves. For instance, the many authors of "Boceprevir for untreated chronic HCV: A Randomized Clinical Trial" try to foreclose these worries by insisting that they have attained a more

[11] For a more detailed analysis of these issues, see K. Brad Wray's contribution to this volume.

traditional authorial status. Their article states, "Each author vouches for the fidelity of the trial conduct to the protocol and the completeness and accuracy of the results and data analyses" (Poordad et al. 2011, 1196). Unfortunately this is radically implausible, if taken as an epistemic claim. The article has 15 authors in four countries employed by 11 academic institutions plus Merck. It is presented on behalf of the SPRINT-2 investigators. The 68-page appendix lists 173 principle investigators from 11 countries who participated in the study, each of whom presumably relied in turn upon a wide variety of support staff to actually implement the study, keep track of the data, and communicate with other participants and with Merck. With the number of people playing some role in the path from study design to implementation to publication running in the hundreds or even thousands, the idea that *each* of the 15 authors could know that *everyone* involved was competent and reliable, that they adhered to the protocol in the same way, that the data set was complete and accurate, and so on is absurd. In radical, distributed biomedical collaborations, testimonial knowledge securing each part of the study is impractical.

5.3 Neither Catch-and-Toss nor Centralized Control

A prima facie response to our argument so far is to admit that radically collaborative science articles, as currently produced, have no author, but also to insist that scientific practice could be reorganized to bring it under the catch-and-toss or the centralized-control models. In that case, climate science, biomedical research, and similar massively collaborative, distributed, interdisciplinary sciences would differ from our original examples only in size and not in deep epistemic structure. We contend, however, that material constraints on this kind of research and the complex normative pressures it is under make it hard to see how to reorganize it in these ways. There are multiple barriers to the effective production of authorship in climate science and biomedical science that are not so easily overcome.

Consider first the attempt to assimilate authorship in radically collaborative science to catch-and-toss authorship. Even if there is no one person who can recreate a study, or even understand all its parts, it is not obvious why this is different from relying on a friend to run your statistics. One response to these worries might be to suggest that we reform the organization of radically collaborative science to allow transparency and coordination of methodological choices and standards. We need to ensure that information about the pressures that shape various parts of the study is readily available, on this line. Furthermore, we need mechanisms that guarantee that each author can have confidence in the competence and honesty of each other author. If we do this, we could, the argument goes, turn radically collaborative research into a huge catch-and-toss project. All the authors could take responsibility for the whole study because they have testimony-based confidence in all its parts and could hold their coauthors accountable for their interest-driven choices if needed. They understand what each of their collaborators contributed, know that each of their collaborators is competent and trustworthy with respect to her part of the whole, and can explain how those parts fit together. Legitimate collaborations would then have an embarrassment of riches: multiple authors, not just one author or no author at all.

In the catch-and-toss model, research can have multiple authors because they understand each part of the project and understand how they fit together, or because they understand why each researcher is trustworthy and accountable when it comes to her own part of the whole. But this requires the ability to know how all of the pieces of the research fit together into a coherent whole. This is not possible in the radical collaborations like those we discussed above. No researcher can vouch for the totality of the decentralized, spatially and temporally distributed epistemic labor. Radical collaborations draw on different kinds of expertise and respond to different kinds of pressures that arise as a result of local context. There is no way for a climate scientist working with CM2.x to know the role of every action, value, and decision that went into building the model; there is no way for any individual to personally vouch that every contributor was competent and reliable in making his choices. Perhaps more importantly, even assuming the competence of everyone involved, there is no one who can know that those choices cohere with the methodological choices made by others who played a role in producing this model. Some of the history of CM2.x reaches back decades and involved people and situations that our hypothetical single climate scientist has never even heard of.

The reasons why radically collaborative research can't be brought under the centralized-control model are perhaps more interesting. Whereas de Prony relied on widely distributed epistemic labor, he was in charge of designing and implementing the project. He established the methodology of the study, relying on information processing that was largely mechanical and that yielded clear data that just needed to be assembled at the end of the production line. De Prony's computers did not need specialized epistemic skills, nor did they need to make their own methodological choices or establish their own epistemic standards; they simply executed the relevant functions. De Prony did not need to trust the expertise of his computers because they were not called upon to exercise any.

In other words, de Prony faced an engineering problem rather than an epistemic problem. As long as there were structural mechanisms in place to guarantee that his computers executed their function sufficiently reliably, and as long as he could count on relatively reliable information flow from the margins back to the center, his reliance on others to carry out the epistemic labor was not interestingly different from relying on any kind of data-collecting or data-storing instrument. De Prony was the only one who was establishing methodological practices and epistemic standards, and he was accountable for the justifiability of the methods and the accuracy of his results (including his method for ensuring the reliability of his "tools"). This is the primary reason why he was in a position to take authorial responsibility for the representation of the relevant results.

Crucially, radical collaborations yield an epistemic problem rather than an engineering problem. In the cases that we have discussed, many different people are engaged in substantive epistemic labor; they are called upon to *use their judgment* and to *establish their methodology*, often in response to real-time messy pressures. This is the main reason why there cannot be someone who ensures the reliability of every decision and who collects and coordinates the information. No one has the relevant expertise to ensure that the epistemic practices and methodological standards used by everyone in the study are reliable. And lack of confidence in *reliability* isn't the only problem;

since they are required to exercise judgment and expertise, the members of a radical collaboration cannot be treated as mere information processors. Their methodological choices and judgments contribute to the overall justificatory story in a way that is lost in simple attempts at collating results. Hence, no matter how much we work on centralizing and coordinating collaborative research projects, and no matter how much we improve transparency and information flow, there is no one who is in a position to play that authorial role of someone like de Prony.

This last point raises a more general and deeper problem: we cannot assume that different disciplines each have fixed methodologies and set epistemic standards. There are reasonable professional disagreements over how to proceed, but in any given situation, methods and standards of justification depend on the goals of the inquirers and the local pressures they face. If there were universally established, fixed methods standards for each type of research, we could set up something similar to the central-control model; we would just need to create structural mechanisms that would guarantee that researchers would properly adhere to these methods and standards, along with reliable methods of information flow and coordination. In this case, the researchers could, in effect, function as *discipline-specific machines* whose results could be interfaced in a modular fashion. But it is precisely because this is not the case that we need researchers who have specialized epistemic skills and judgment; this is the reason why no central coordinator can establish in advance what everyone's proper standards should be. We mentioned that these sorts of problem arise in climate science earlier, but to make the generality of this issue clear, it will help to consider two additional examples.

First, it is standard for the principal investigators on a clinical trial to employ a dedicated statistician. Often, in an industry-funded trial, the statistician will be an industry employee. The lead authors may or may not be in a position to check over the work of the statistician, looking for glaring errors. But the whole reason why research projects bring on a statistician, rather than using some sort of statistics software, is that there are substantive, nontrivial choices to be made among different kinds of statistical analyses. Running the statistics on a study is not merely a mechanical procedure, but rather requires subtle sensitivities to the project and pressures at hand. The PIs don't have the skills to make these subtle judgment calls, or they would just do the analysis themselves. But when research is radically collaborative, there is no clear way to guarantee that the goals that shape even a totally competent statistician's choice of methods and standards coordinate with the goals of other researchers. This is especially clear (but not only so) in the case of industry-employed statisticians working on privately funded studies, who have a clear motive to choose a method of analysis that is favorable from a marketing perspective. Authors often have no mechanisms for ensuring that the goals of the statistician, and her understanding of the questions and pressures at stake in the research, coordinate with those of other researchers.

A different kind of example comes from local pressures that are opaque to spatially distributed collaborators. Consider an investigator who is on the ground collecting data on vaginal macrobicide and HIV transmission in a remote African village. It is likely that she will need to make many judgment calls as to whether a given subject was compliant enough with the protocol or providing reliable enough reports to be included in the study. But this is tricky business where there are cultural barriers, populations with low scientific literacy, and cultural forces and power relations that impede free and informed talk about sexual behavior. There is no single right answer that can be mechanically generated as to when a subject's data should be included or not; this is a judgment call. Furthermore, it is a judgment call that will be shaped by a variety of unpredictable factors, such as how hard it is to retain subjects in the study, how dire the communication difficulties are, the time frame for the funding, and so forth. The PIs back in North America will never be in a position to dictate these standards in advance or to know which standards were used in each case. Crucially, ensuring honesty, competence, and the free flow of information will not solve this problem.

In sum, since different researchers are working in different locations, using different disciplinary expertise, are driven by different working understandings of the goals of the research, and subject to different local pressures resulting in varying standards, there seems to be no built-in guarantee that the justificatory stories that undergird the various pieces of the study together form one coherent justification, for which the group as a whole can be accountable. If the represented results are challenged, there may be *no* single justificatory story to be told about the methodological choices made and the epistemic standards used—not even one that would need to be told piecemeal by the various participants. Thus it seems like there is no reason to believe that the group collectively can be accountable for the finished product. Hence there seems to be little reason to count them as forming a group author.

Before moving on we will briefly consider a different way of bringing radically collaborative research under the centralized-control model. We develop this alternative and argue against it in much more detail in a different paper (Winsberg, Huebner, & Kukla 2014). Perhaps the central controller need not control each part of the research process and methodology, but instead the flow of information itself. That is, she might implement a social model that is designed to guarantee that the input from different researchers, operating according to different standards and in light of different pressures and interests, coheres into a reliable whole because of how it is combined. The reliability of the results would be defended not in terms of the specifics of the methodology of each part, but in terms of the reliability of the system for combining them—in something like the way that we can count on the "wisdom of crowds" even (and indeed particularly) when we have no reason to count on the wisdom of individual crowd members.

We cannot address this suggestion in detail here, but notice two things. First, as things currently stand, radically collaborative sciences like climate science and biomedicine have nothing resembling such a social model. They have not even acknowledged that there is an epistemic issue concerning how the different parts of such research projects combine. Second, whoever was in control of such a project would have to have an explicit commitment to her social model and a justification of its reliability. If we are interested in accountable authorship, then it is not enough for us, from the outside, to decide that a result is reliable despite its inner chaos, perhaps for wisdom of crowds sorts of reasons. Rather, the central controller would count as epistemically accountable only insofar as she was committed to the reliability of such a social model and ready to defend it or give it up as needed. In any case, if we think of *authors* as primarily authoring a system for coordinating parts whose individual reliability cannot be defended, then we are radically changing our conception of scientific authorship.

5.4 How Deep Is the "No Shared Epistemic Standards" Problem?

In the preceding section, we talked about the problem of there being no shared epistemic standards that can be used to (more or less mechanically) assess the methodological choices of various collaborators. Broadly speaking, there are three things we could mean by "shared epistemic standards."

First, we might be making the familiar claim that there is no logic of induction. It is now virtually a truism in the philosophy of science that there exists no single articulable set of rules that can deliver, mechanically, probabilities for a hypothesis given a specification of the evidence available in favor of (and against) it. This is a sense in which there are "no shared epistemic standards": if there are rules for the epistemology of science, they are "one off." There might be, in other words, rules of thumb that can sanction empirical inferences in particular restricted domains, but these rules are brittle, and they do not travel well. They also, it is often argued, involve tacit knowledge that involves years of training to acquire.

This kind of lack of shared standards poses obvious difficulties for the sorts of strategies discussed in the last section, but they are not obviously insurmountable. Even if epistemic standards are fragmented and difficult to communicate, it might, in principle be possible for some centralized authority to master them. So the argument against easy answers to the author problem from this kind of failure of shared standards exists but is relatively weak.

A second way of fleshing out the claim that there exist no shared epistemic standards would be to maintain that even given a particular epistemic task, a hypothesis, a body of background knowledge, and a body of evidence, there exists no uniquely correct answer concerning whether the evidence sufficiently supports the hypothesis. This version of the "no shared standards claim" is full-blown epistemic relativism. We reject this, or at the very least, we recognize that such relativism is incompatible with our notion of authorship, which centrally involves the concept of epistemic accountability. Genuine accountability involves being able to call the accountable person onto the mat and ask if the relevant standards have been met; the very ideas of accountability and authorship makes no sense if there are literally no common standards at all. I cannot be held accountable for my choice of ice cream flavor.

But there is a third notion of "shared epistemic standards" whose failure poses a more serious and specific problem for radically collaborative research than the mere absence of a logic of induction. Note that even if there were a logic of induction, no scientific hypothesis would ever be established with certainty on the basis of a finite amount of evidence. The judgment whether a hypothesis should be accepted in light of the evidence involves two kinds of trade-offs. First, let us call the capacity of an investigation to generate persuasive results given a certain amount of effort and resources and data collection, the "power" of the investigation. Methodological choices involve a trade-off between the power and the reliability of the investigation. Second, even given a choice along that continuum, methodological choices involve a further trade-off between a desire to avoid false positive conclusions and a desire to avoid false negative conclusions—a so-called balance of inductive risks (Churchman 1948; Rudner 1953; Douglas 2000, 2004, 2009; Wilholt 2009). Decisions about how sure we need to be to accept a hypothesis always depend on the seriousness of making a mistake. Thus there cannot be any general or objectively correct answer to how such a trade-off should be balanced. Values and interests inevitably play an important role in determining the "seriousness" of accepting false positives and rejecting false negatives: to use Richard Rudner's classic example, our concern with false negatives will be much higher if we are studying whether a drug has a potentially lethal side effect than if we are testing for defects in a set of belt buckles (Rudner 1953).

Since scientific inference is always uncertain, there is no avoiding making a judgment about how to balance inductive risks, and particular interests and investments will often affect this judgment: a drug company may well set a lower standard for "establishing" efficacy than a consumer group. Furthermore, as Douglas (2000, 2009) and others have argued, such interest-dependent judgments occur throughout the research process. How data are classified and coded, which sorts of screening tests are used, which methods are employed in smoothing and correcting data, and indefinitely many other judgments involve this type of inductive risk balancing. For example, whether researchers classify slides of rat tumors as benign or malignant turns out to depend, to a very large degree, on the goal of the study and their stake in the outcome (Douglas 2004).

In such situations, there is no value-free notion of a *correct choice* of distribution of inductive risks, since *any* choice will involve trade-offs. Likewise there is no value-free notion of the "correct epistemic standard," even in the "one-off" sense discussed above.[12] But remember where this leaves us: not only do different researchers in a radically collaborative research project use different methodological standards and make different choices, and not only do we *need* them to be able to exercise their judgment in this way, but there is also no *right answer* as to how they should make these choices—no universal set of principles by which their performance can be measured. Each researcher faces micro- and macro-pressures that yield a local, one-off set of interests and values, and these will inevitably shape her choices all through the research process. Not only can we not typically recover these interests and choices (which, as we have pointed out, are likely often opaque to the researchers themselves), but even if we could, we would have no stable meta-standard for measuring their scientific reliability. Indeed, the relevant notion of reliability is at the center of what requires individual, value-laden judgment.

What sorts of interests and values might make a relevant difference to methodological choices, including inductive risk balancing in the context of radically collaborative research? Climate science and biomedicine are normatively rich sciences that are driven by social values and demands in obvious ways. We have direct practical interests

[12] This is not the same thing as full-blown epistemic relativism, since we can each still offer *reasons* for having chosen one methodological option over another.

in the results of these sciences—interests that are personal, ethical, political, and economic. Different stakeholders obviously care in various ways whether a climate model gives encouraging or alarming results, whether a drug is more effective than the current standard of care, and so forth. A great deal of attention has been devoted to exploring how profit motives shape industry-funded clinical drug trials, and it may be possible to address the systematic structuring of studies by clear, recoverable interests (cf., Healy & Cattell 2003; Elliott 2011a, 2011b; Kahn 2004, 2006; Sankar & Kahn 2005; Sismondo 2007, 2008, 2009). But what matters more for our purposes is the micro-interests and micro-values that shape individual researchers' local practices, rather than the large-scale social investments that shape the research project as a whole and might be relatively easier to uncover and critically assess.

When bioethicists and philosophers of science have worried about the role of values and interests in shaping methodological choice points in biomedical research, they have focused overwhelmingly on *financial* interests. Indeed, concerns over the influence of financial pressures are justified. Private industry plays a huge and ever-increasing role in funding biomedical research. Scandals over ghostwriting, selective data use, selective publication, and ad hoc study designs are common, and often focus on the role of pharmaceutical companies and for-profit companies in controlling both the methods and the message of the science they fund. These scandals typically reveal a top-down, organized attempt to shape publications in ways that will increase the profits of a specific stakeholder. But published studies in biomedical science typically disclose dozens or even hundreds of conflicting financial interests. Collaborators often receive funding and support in the form of shares, grants, and lecturing fees, among many others. Collaborators from one part of a study are unlikely to know all the micro-pressures that other participants are under to secure grants, satisfy donors, or wrap up a study.

Furthermore, financial interests are only one kind of pressure that shapes the methodological decisions and standards that are at play in biomedical research. At any given site, postdocs might be under pressure to please their advisers, there might be scientists who are attached to pet hypotheses or concerned to build their reputations, faculty might be under pressure to flesh out their annual reports or make tenure, and there may be disciplinary turf wars and grand ideological battles that guide the research. This is compounded by professional competitions and investments, along with pressures from local universities and communities. The course of the research is shaped by a wide variety of micro-interests, even where everyone behaves honestly and nothing is hidden. These interests are not typically coordinated with one another; often they may not be consciously available even to those that have them.

One source of such micro-interests—which bioethicists have discussed at length—derives from the fact that most collaborators in clinical research are healthcare professionals, so they come into the research project inhabiting dual roles with potentially conflicting duties. They are professional caregivers and healers, as well as scientists. But notice that these dual roles can create pressures that can affect inductive risk balancing and other methodological choices. For example, when faced with a subject who has a medical need, it is not obvious that the duty to stick closely to a scientific protocol will trump the duty to care for a person. In fact, Charles Lidz and his colleagues (2009) report that 64% of clinical researchers thought that researchers *should* depart from protocol to improve medical care, and many of them reported giving restricted medication to subjects, recruiting subjects they knew to be ineligible for a trial, and keeping subjects in a trial after they met the termination criteria to improve their health. As they note, such deviations may significantly affect the results of clinical trials.

5.5 Structured Radical Collaboration in High-Energy Physics

Up to here, we have focused on radically collaborative research in climate science and biomedicine. These are sciences in which it is not surprising that values and interests play a substantive role, given the socially pressing, normatively complex nature of their missions. Perhaps, then, our worries are not really about authorship in radical collaboration per se, but rather concern value-rich science. In this section we consider a "pure" science that still operates through radical collaboration by necessity. Research in high-energy physics (hereafter HEP) must be distributed across numerous laboratories (which are often located in different locations throughout the world). In the case of the Organisation Européenne pour la Recherche Nucléaire (CERN), publications reporting collaborations on a single experiment may include as many as 1,000 authors, listed in alphabetical order without regard to seniority.[13] Like climate modeling and biomedical research, research in HEP relies on technological and symbolic resources that are rarely shared among different labs. The size and complexity of the detectors, the excessively long duration of the experiments (some lasting as long as 20 years), as well as the degree of collaboration required to produce a result, make isolated authorship impossible. But HEP is not clouded by policy entanglements, and, indeed, it seems to have little if any foreseeable practical import. Thus HEP is a good candidate for a science that may be radically collaborative without raising worries about the structuring role of multiple interests.

But consider an anecdote that was relayed to us. There were two major groups looking for the Higgs particle at CERN: ATLAS and CMS. When ATLAS reported data consistent with the observation of the Higgs particle at 2.8 sigma, the buzz around CMS was that they needed to do whatever was necessary to "boost their signal." This meant shifting their distribution of inductive risks to prevent them from falling too far behind the ATLAS group—toward higher power, at the expense of reliability, or toward a lower probability of missing a Higgs-type event, at the expense of a higher probability of finding a false positive. Hence even here, it seems that we see the influence of local pressures and interests on methodology.

It may seem that such standards could be chosen collectively, or managed centrally, in ways that preserve the methodological transparency that is necessary for collective

[13] The Stanford Linear Detector (SLD) worked to maintain "the coherence of the group and the *de facto* recognition that contributions to a piece of physics are made by all collaborators in different ways" (Galison 2003, 332). Each physicist was listed as a coauthor, and that the first publication on a particular topic included engineers as well. Authorship was not limited to those who "contributed" to the reported result, but included everyone who had worked with a research group for a year (cf. Galison 2003, 334?335).

authorship. But a little bit of background should disabuse us of such a fantasy. The LHC detectors generate huge amounts of raw data: approximately 23 petabytes/second of raw data.[14] Hence, a triggering mechanism must be used to decide which data to keep and which to ignore. Strikingly, this triggering process involves human judgment. When collisions occur, one cannot simply see a Higgs particle. At best, one sees the spray of particles that a Higgs event would produce if it occurred. But we are unable to calculate from first principles what this spray will look like because we don't have a good enough theory of the strong interaction. So the calculation has to be semi-empirical, and some of the relevant empirical facts have to be observed on the fly. Very time-sensitive judgments have to be made, and they have to be made by widely distributed agents. And, of course, every judgment involves a distribution of inductive risks. Thus it is hard to know how a perception that "we are falling behind the ATLAS group in the race" will affect the judgments on the ground.

This is something that people at CERN implicitly understand. They attempt to manage the role that interests might play in shaping perception by, for instance, using blinding procedures: The people making the empirical judgments about whether a Higgs event has occurred cannot see the part of the background information that is relevant to the calculation of whether a "trigger" will be reached when they make their judgment. This separates the roles of those who decide how much to boost the signal from those who evaluate the resulting patterns; in other words, this structural mechanism is designed to minimize any distortion issuing from the interests of the scientists. Whether or not this technique helps address this particular problem, it indicates a way in which inductive risk balancing continues to occur in unpredictable and, perhaps, unrecoverable ways throughout the research process, even where the research does not aim at some obviously value-laden goal.

Nonetheless, we think there are two relevant differences between HEP and our other two cases that make us worry less about the reliability of the results of HEP. (1) In HEP, the distribution of inductive risks is linear, in the sense that the Higgs Boson either exists or it does not. Compare with the climate case, where there are a wide variety of desirable prediction successes: global mean surface temperature, precipitation, sea level, ice melt, drought, storms, and other regional effect, and so forth. These are inductive risk considerations that pull in many directions; researchers can have any of a wide, multidimensional array of investments in various outcomes and parameters at both the micro- and the macro-level. In HEP, there is only one dimension of pull: you are either erring on the side of false positives or false negatives when it comes to finding the Higgs Boson. This would seem to make the problem of "conflicting standards" much more manageable, at least in principle. (2) Relatedly, in HEP, while there is a distribution of inductive risks at any given moment in time, in the long run, there will not be one. As more and more data are collected by the LHC, the power of the experiment gets larger and larger, and the balance of DIR gets smaller and smaller, until eventually, the choices will not matter. Likewise, the phenomena in HEP do not themselves change, so any attempts to understand them are, at least in principle, both cumulative and reproducible. In contrast, in climate science and biomedical research,

[14] Approximately 25 megabytes per event, times 23 events per beam crossing, times 40 million beam crossings per second in the center of the detector.

the central questions are fundamentally time sensitive, both because the research questions are inherently local and the phenomena under scrutiny themselves dynamic. Of course, by 2100, we will know with certainty what the 2100 climate looks like. And in 2100, we will almost certainly retroactively know how effective some particular drug was in some population. But our biomedical research practices are, of necessity, set up to give us answers *now*. HEP is set up to give us good results as soon as possible, but the data will continue to trickle in until we are, for all intents and purposes, sure of their implications.

While (at least in the long run) we may be less concerned about the *reliability* of HEP than of climate science or biomedicine, we have not found great reason to be comforted about the prospects for traditional authorship in this domain. Of necessity, a widely distributed, massive group of highly specialized researchers must participate in these projects. These researchers are not in a position to either understand all their colleagues' contributions or to track and vouch for their local choices and judgments, and it still seems unlikely that this problem will be eliminated by trying to make the science more mechanistic or transparent.[15] That we can be more confident that HEP will converge on reliable conclusions eventually is politically comforting, perhaps, but it does not help give us a model of authorship in radically collaborative research that should comfort the social epistemologist.

5.6 Conclusion

We have argued that there are—at present—no plausible strategies for understanding or ensuring accountable authorship in the context of radical collaborative research. We've tried to show that there is no one who is in a position to know, understand, or explain all of the epistemically significant decisions that have been made in the production of a purported piece of scientific knowledge. In radical collaborations, catch-and toss authorship cannot be sustained because the individuals who are involved in a research project cannot recover all of the epistemically significant decisions that have been made along the way; similarly, the necessity of local judgments that are made on the fly precludes the possibility of top-down, centralized control over knowledge production. Finally, as we suggested in the last section, the necessity of value-driven risk-balancing decisions—which show up even in a "pure" science like HEP—make genuine *group authorship* difficult to sustain. Together, these arguments suggest that neither individual nor multiple nor group authorship is plausible in radically collaborative research, at least in anything like its current organizational form. As things stand, authorship *simpliciter* remains impossible in most radical collaborations.

We have focused on the ways in which these issues arise in climate science and biomedical research. This is because the data collected and reported in these sciences are epistemically, morally, and politically important. Scientists working in these areas confront issues that are both socially pressing and normatively loaded, and it matters

[15] Indeed, high-energy physicists have been exceptionally interested in authorship questions and have developed various techniques for regulating and streamlining the authorship process; see footnote 13. But these have not been targeted at the kinds of concerns we have focused on here.

deeply that the data they reported are accurate and that there is someone or some people who can take responsibility for this accuracy. This is not merely because we want to get the facts right, but also because we often need to decide—right now—whether to remove a particular drug from the shelves or whether to implement a policy that can sustain public health in the face of global climate change. The ability to interpret the results reported in radically collaborative publications, and to decide when and how to trust these results, matters not just epistemologically but also morally and politically.

Until now, radically collaborative science *as such* has not been identified as a distinctive phenomenon with its own epistemological and ethical challenges and complications. But such collaborations dominate many of the most socially pressing areas of science. This form of science is increasingly well represented in top journals, and it is receiving more and more funding. We contend that the existing models of collective authorship and collaboration are unequipped to deal with the complexities that arise when epistemically robust values and judgments must be made at many disparate points in the production of radically collaborative data.

We also maintain that the models of collaboration and authorship that are applicable to small groups, where everyone can understand the judgments and decisions made by others, cannot simply be scaled up to address the issues that we have addressed. Importantly, this is not because we don't yet know how to scale them up. It is because scaling up is not the right way to address the issues we have raised here. These form a novel class of epistemic and ethical challenges that arise specifically in sciences of scale. They are not, even in principle, the kinds of issues that we could scale up simpler models to address; they originate in the messiness that is produced by massively complex, distributed research architectures. Social epistemologists need to think creatively about authorship and accountability in the context of radically collaborative research, in which the production of a scientific result requires multiple, disparate kinds of experts, widely spread out across various locations, working under time- and context-specific conditions.

References

Andersen, H. and S. Wagenknecht. (2013). "Epistemic dependence in interdisciplinary groups." *Synthese* 190: 1881–1898.

Biddle, J., and E. Winsberg. (2009). "Value judgments and the estimation of uncertainty in climate modeling." In *New waves in the philosophy of science*, ed. P. D. Magnus and Jacob Busch, 172–197. New York: Palgrave MacMillan.

Clark, A. (1998). *Being there: Putting brain, body, and world back together*. Cambridge: MIT Press.

Churchman, C. W. (1948). *Theory of experimental inference*. New York: Macmillan.

Douglas, H. (2000). "Inductive risk and values in science." *Philosophy of Science* 67 (4): 559–579.

Douglas, H. (2004). "Prediction, explanation, and dioxin biochemistry: Science in public policy." *Foundations of Chemistry* 6 (1): 49–63.

Douglas, H. (2009). *Science, policy, and the value-free ideal*. Pittsburgh: University of Pittsburgh Press.

Dunne, J. (2006). "Towards earth system modeling: Bringing GFDL to life." Presented at the ACCESS 2006 BMRC Workshop. http://goo.gl/lPnwz (accessed Jan. 11, 2011).

Elliott, K. (2011a). *Is a little pollution good for you? Incorporating societal values in environmental research*. New York: Oxford University Press.

Elliott, K. (2011b). "Direct and indirect roles for values in science." *Philosophy of Science* 78: 303–324.
Galison, P. (2003). "The collective author." In *Scientific authorship: Credit and intellectual property in science*, ed. Mario Biagioli and Peter Galison, 325–353. London: Routledge.
Gilbert, M. (1989). *On social facts*. London: Routledge.
Gilbert, M. (1996). *Living together: Rationality, sociality, and obligation*. Lanham, MD: Rowman and Littlefield.
Grier, D.A. (2005). *When computers were human*. Princeton: Princeton University Press.
Gowers, T., and M. Nielsen. (2009). "Massively collaborative mathematics." *Nature*, 461 (15): 879–881.
Healy, D., and D. Cattell. (2003). "Interface between authorship, industry and science in the domain of therapeutics." *British Journal of Psychiatry* 183: 22–27.
Huebner, B. (2013). *Macrocognition: Distributed minds and collective intentionality*. New York: Oxford University Press.
Kahn, J. (2004). "How a drug becomes 'ethnic': Law, commerce, and the production of racial categories in medicine." *Yale Journal of Health Policy, Law, and Ethics* 1: 1–46.
Kahn, J. (2006). "Race, pharmacogenomics, and marketing: Putting BiDil in context." *American Journal of Bioethics* 6 (5): W1–W5.
Kukla, R. (2012). "Author TBD: Radical collaboration in contemporary biomedical research." *Philosophy of Science* 79 (5): 845–858.
Lenhard, J., and E. Winsberg. (2010). "Holism, entrenchment, and the future of climate model pluralism." *Studies in the History and Philosophy of Modern Physics* 41 (3): 253–262.
Lidz et al. (2009). "Competing commitments in clinical trials." *IRB: Ethics and Human Research* 31 (5): 1–6.
List, C., and P. Pettit. (2011). *Group agency: The possibility, design, and status of corporate agents*. New York: Oxford University Press.
Poordad, F., et al. for the SPRINT-2 Investigators. (2011). "Boceprevir for untreated chronic HCV: A randomized clinical trial." *New England Journal of Medicine* 363: 1195–1206.
Rolin, K. (2008). "Science as collective knowledge." *Cognitive Systems Research* 9 (1–2): 115–124.
Rudner, R. (1953). "The scientist qua scientist makes value judgments." *Philosophy of Science* 20 (1): 1–6.
Sankar P., and J. Kahn. (2005). "BiDil: Race medicine or race marketing?" *Health Affairs* W5: 455–463.
Shackley, S., J. Risbey, P. Stone, and B. Wynne. (1999). "Adjusting to policy expectations in climate change modeling: An interdisciplinary study of flux adjustments in coupled atmosphere-ocean general circulation models." *Climatic Change* 43 (2): 413–454.
Shieber, J. (Forthcoming). "Toward a truly social epistemology: Babbage, the division of mental labor, and the possibility of socially distributed warrant." *Philosophy & Phenomenological Research*.
Sismondo, S. (2007). "Ghost management: How much of the medical literature is shaped behind the scenes by the pharmaceutical industry?" *PLoS Medicine* 4 (9): e286.
Sismondo, S. (2008). "How pharmaceutical industry funding affects trial outcomes: causal structures and responses." *Social Science and Medicine* 66 (9): 1909–1914.
Sismondo, S. (2009). "Ghosts in the machine: Publication planning in the medical sciences." *Social Studies of Science* 39 (2): 171–198.
Thagard, P. (2002). "Collaborative knowledge." *Noûs* 31 (2): 242–261.
Tuomela, R. (2002). *The philosophy of social practices: A collective acceptance view*. Cambridge: Cambridge University Press.
Weart, S. (2010). "The development of general circulation models of climate." *Studies in History and Philosophy of Science Part B* 41(3): 208–217.
Wilholt, T. (2009). "Bias and values in scientific research." *Studies in History and Philosophy of Science* 40 (1): 92–101.
Winsberg, E. (2006). "Models of success vs. the success of models: Reliability without truth." *Synthese* 152: 1–19.

Winsberg, E. (2010). *Science in the age of computer simulation*. Chicago: University of Chicago Press.
Winsberg, E. (2012). "Values and uncertainties in the predictions of global climate models." *Kennedy Institute of Ethics Journal* 22 (2): 111–137.
Winsberg, E., B. Huebner, and R. Kukla. (2014). "Accountability, values, and social modeling in radically collaborative research." *Studies in History and Philosophy of Science: Part A* 46: 16–23.
Wray, K. (2002). "The epistemic significance of collaborative research." *Philosophy of Science* 69 (1): 150–168.
Wray, K. (2006). "Scientific authorship in the age of collaborative research." *Studies in History and Philosophy of Science Part A* 37 (3): 505–514.
Wray, K. (This volume). "The impact of collaboration on the epistemic cultures of science."

6

The Impact of Collaboration on the Epistemic Cultures of Science

K. BRAD WRAY

My aim is to examine how collaborative research affects the epistemic cultures of science. Central to my analysis is the notion that the views expressed in the coauthored articles that research teams produce represent the views of the group as a whole rather than the views of the individual group members. I begin by arguing that some groups of scientists hold views that are irreducibly the views of the group. It is the group that is properly described as *collectively* holding the view. Specifically, I claim that research teams often hold irreducibly collective views. And some other scientific groups have this capacity as well.

I then consider some normative issues that have arisen in the epistemic cultures of science, issues specifically related to collaborative research. First, I examine how collaborative research threatens to erode the traditional notion of authorship in science. Specifically, I will examine (i) problems arising in Big Science, where articles are sometimes authored by hundreds of scientists, and (ii) the challenges research communities face when they attempt to craft authorship criteria that specify the conditions under which one has a legitimate claim to authorship in a collaborative research project. Second, I examine how collaborative research in science affects refereeing. I argue that there is reason to believe that journal editors and journal editorial boards are out of step with the changes that have occurred in the cultures of science as collaborative research has become more common. Thus, I argue that the norms of authorship and the refereeing practices have not developed to address the challenges encountered in the cultures of science where collaborative research is commonplace.

Before proceeding some prefatory remarks are in order. I use the term "epistemic culture" in the same way that sociologists of science, like Karin Knorr Cetina, use the term. Thus, I regard as "epistemic cultures … cultures that create and warrant knowledge" (see Knorr Cetina 1999, 1). My focus will be narrowly on scientific epistemic cultures, and I assume that the scientific specialty is a key unit of analysis. I do, though, recognize that much research, especially in Big Science, cuts across specialty and disciplinary lines. The term "Big Science" denotes the "large-scale character of modern science" (see Price 1963, 2). The term was popularized by Derek de

Boyer-Kassem, Mayo-Wilson, & Weisberg. *Scientific Collaboration and Collective Knowledge*.

Solla Price, and it was meant to highlight the shift in the epistemic cultures of science associated with the large research facilities that are state supported which have become increasingly popular since the Manhattan Project.

6.1 Collective Knowledge and Science: A Defense

Margaret Gilbert (2000) was one of the first to argue that the phenomenon of collective belief was relevant to science. She argues that a scientific specialty can hold a view that is aptly characterized as irreducibly the view of the specialty community. When a view is irreducibly the view of the group, it is the group as a unit that holds the view rather than the individual members of the group. Thus, for example, on Gilbert's account when gastroenterologists came to believe that bacteria are the cause of many stomach ulcers, this belief was held collectively, by the community of gastroenterologists, as opposed to merely being held by many individual gastroenterologists. So in order to bring about this change, to make this belief about the cause of ulcers the accepted view among gastroenterologists, the view of the research community had to be changed. And this involved more than merely changing the views of many individual gastroenterologists. Gilbert refers to groups that hold irreducibly collective views as plural subjects, and her account is often referred to as the plural subject account of collective belief.

Though I agree with Gilbert's general claim that groups of scientists are capable of holding views that are irreducibly the views of the group, elsewhere I have raised two concerns about specific aspects of her account. It is worth reviewing these concerns and my suggested amendments, as they are assumed throughout the remainder of the chapter.

First, elsewhere I have taken issue with Gilbert's claim that the cognitive attitudes of a group are aptly characterized as beliefs (see Wray 2001). Instead, I argue that the cognitive attitude of such groups is better characterized as acceptance, where acceptance is distinct from belief. When a group accepts a view or proposition, they do so for pragmatic reasons, and they do so voluntarily. Beliefs, on the other hand, are acquired involuntarily, and one is compelled to believe what one thinks is true.[1] Groups can and *often* do accept views that the various members of the group individually believe to be true. Groups are often inclined to do so because accepting and acting on true claims is often an effective means to pursuing their goals. The views a group or individual accepts are used as guides to actions, and true claims are often effective guides to action.[2] Still, in principle, the view of a group can differ from the views

[1] See Kay Mathiesen (2007, Table 1) for a full list of the features identified as distinguishing belief from acceptance. As Mathiesen notes, the distinction is the focus of the debate between Rejectionists, who claim that groups cannot have beliefs, and Believers, who claim otherwise.

[2] Don Fallis (2007) has a useful analysis of how pragmatic considerations need not undermine our pursuit of the truth. Rightly, he notes that pragmatic considerations can be helpful incentives for getting at the truth (see Fallis 2007, 270–271). "Good epistemic outcomes tend to be good outcomes all things considered" (Fallis 2007, 275). Kent Staley (2007) also argues that pragmatic considerations can advance scientists' epistemic goals.

of the individual members of the group. And the most interesting cases of collective acceptance are those in which there is a discrepancy.

In light of this amendment to Gilbert's plural subject account of collective knowledge, when ascribing collective knowledge to groups we need to broaden the definition of knowledge. Knowledge becomes justified true *belief or* justified true *acceptance* (see Wray 2007a, Section 2). This more liberal conception of knowledge was proposed earlier by L. Jonathan Cohen, though he was not specifically concerned with the cognitive states of groups (see Cohen 1992, 82). Rather, Cohen insisted that the conception of knowledge relevant to scientific investigations of the laws of nature involves acceptance, whereas the conception of knowledge relevant to experimental data involves belief.

Second, contrary to what Gilbert claims, I also argued that *scientific specialties* and *the scientific community* as a whole lack the capacity to hold views that are irreducibly the views of the group. I did, though, argue that *research teams* are capable of holding views that are irreducibly the views of the group (Wray 2006; 2007a). What distinguishes research teams from the former sorts of groups is that research teams have organic solidarity (Wray 2007a).[3] The notion of organic solidarity is adopted from Émile Durkheim (see Durkheim 1893/1984, 84–85). A group has organic solidarity when it is organized by a division of labor where the various members depend upon each other to fulfill their role in realizing the goals of the group. The various members of the group function as organs in an organism. Laboratory teams and other scientific research teams are typically organized by a division of cognitive labor (see Knorr Cetina 1999; Thagard 1999). Scientific specialties, on the other hand, merely have mechanical solidarity. The members of a scientific specialty are like-minded only insofar as they have the same beliefs. Their shared beliefs are their *individual* beliefs.[4] A single social group can have both mechanical and organic solidarity. Durkheim recognized this. But the distinction alerts us to the significant differences between social groups that are characterized by a division of labor and function as a unit, and other more inchoate sorts of social groups whose members are similar to each other but lack the cohesion of a group with organic solidarity.

[3] Others have raised criticisms different from the ones I discuss here or suggested refinements to Gilbert's plural subject account of collective belief. See, for example, Alban Bouvier (2004, 2010) and Beatty and Moore (2010). Both Beatty and Moore (2010) and Bouvier (2010) suggest that Gilbert fails to distinguish between types of groups that are genuinely different. Some groups, they note, have formal voting mechanisms in place to determine the view of the group, whereas other types of groups determine the view of the group in a more informal and inchoate manner. Beatty and Moore (2010, 210) introduce the notion of deliberative acceptance to characterize the type of collective acceptance that results from a vote among equals that follows a process of deliberation. Bouvier (2010) recommends making even more distinctions between the various phenomena that Gilbert regards as instances of collective beliefs.

[4] Kristina Rolin (2008) takes issue with my claim that scientific specialty communities cannot have collective knowledge. And Hanne Andersen (2010) has taken issue with aspects of both my view and Rolin's view of collective acceptance in science. Andersen grants that scientific research teams are organized by a functional division of labor. But she insists that research teams have a more complicated structure than Rolin and I recognize. Hyundeuk Cheon (2014) and Chris Dragos (2016) also have weighed in on the debate between Rolin and me about which sorts of groups in science can hold views that are irreducibly the view of the group.

Other groups involved in science besides collaborative research teams are capable of holding irreducibly collective views. For example, committees of scientists convened to determine the official view of the committee are also capable of holding collective views. John Beatty discusses an example of a committee of scientists commissioned with the responsibility of making a recommendation in the determination of a government policy on an issue that depended upon scientific knowledge (see Beatty 2006). And Paul Thagard (1999, Chapter 12) discusses examples of committees involving scientists and laypeople commissioned to determine policies pertaining to health-related issues that depend upon current medical research. These consensus conferences or committees are capable of holding views that are not reducible to the views of the individual members of the group. I will discuss Beatty's example in more detail shortly.

Importantly, scientific research teams not only have the capacities to hold views and to have knowledge. Working collaboratively a group of scientists are also able to pool knowledge from different sources and draw inferences that individually no single scientist in the group working alone would be *justified* in drawing.[5] And there is a substantial body of empirical evidence that suggests that such groups are more effective at reasoning than their constitutive individual members (see Trognon and Batt 2012: Unpublished). Collaborative research thus promises to open up areas of research that no individual scientist could hope to address working alone.

In a recent article, Melinda Bonnie Fagan (2011) challenges the ascription of irreducibly collective knowledge to scientific groups and asks the following question: What do appeals to collective knowledge aim to explain? It is worth answering Fagan's question.

There are three things that appeals to the notion of collective acceptance explain. First, collective acceptance can account for the discrepancies between the beliefs of individual scientists working as part of a research team and the published views of the research team. That such discrepancies exist seems undeniable. In fact, sometimes some of the individual members of a research team will not even fully understand the implications of the articles they coauthor. Reflecting on the nature of collaborative research, one realizes that this is to be expected. One of the chief reasons scientists collaborate with other scientists is to pursue research projects for which they individually lack the relevant background knowledge and skills. By working collaboratively, scientists can accomplish things they cannot accomplish alone (see Thagard 1999, Chapter 6, for an example of this). And even when an individual understands the implications of what the group reports in an article, the individual may not personally endorse or believe all the claims reported. Collective acceptance figures in explanations of the cognitive attitudes of the group when they are distinct from the cognitive attitudes of the individual members of the group. In fact, a group of scientists can endorse a view or set of claims that no individual member of the group wholly endorses.[6]

[5] There is a growing body of literature on *group justification*. See, for example, Schmitt (1994), Rolin (2010), Staley (2010), and de Ridder (2014).

[6] Knorr Cetina (1999) makes the radical proposal that in contemporary science, especially in high-energy physics, it is the experiment itself that is the knowing agent, not the scientists. Ron

Let us consider a concrete example where scientists did in fact act in the manner just described. This example does not involve a research team, but rather involves a committee of scientists. Beatty provides an interesting analysis of how a group of scientists collectively endorsed a report that included many claims that a number of the individual members of the group did not believe or endorse personally. This case involved a number of scientists determining the safe level of radiation to which humans could be exposed (see Beatty 2006). The scientists were commissioned to this task by a government agency, the National Academies of Science (NAS), in the United States in the 1950s. The radiation levels that *individual* scientists believed were safe ranged widely (Beatty 2006, 63; see the table). But the government agency was interested in issuing a policy, and it was not going to await further research. Ultimately, the group of scientists commissioned with the task settled on a view that would represent the view of the group. It was their collective view. Few of the scientists involved would identify the view of the group as their individual view, even after they participated in the exercise. Still, the view stood as the view of the group. In this case, there were pragmatic concerns motivating the individual scientists to participate in the process despite their misgivings about it. Many of the scientists felt that if they did not participate on the committee and make a recommendation about the safe level of radiation, someone else would be assigned the task and that someone else may have been less qualified to do the job.[7]

Similarly, when a research team publishes a coauthored article, the members of the team will often have to compromise. One may have to accept qualifications on the interpretation of data that one does not personally think are necessary. Or one may have to tolerate interpretations that one personally believes exceed what is warranted by the evidence. But to resist compromising is to delay publication, which in turn risks allowing another research team to make and report the discovery first.

Second, collective acceptance can account for the attitudes that individual scientists have while working as part of a research team. Individual scientists who work as part of a research team sometimes express fears that they will not get adequate credit for their contributions to the group's research project (see Wray 2006). They are not necessarily concerned that a particular individual is stealing the credit. Rather the concern is that

Giere (2002) has taken issue with Knorr Cetina's suggestion. Giere prefers to characterize the cognition of research teams as a form of distributed cognition, and maintains that only individual scientists are capable of knowing. Alternatively, we might construe the published article as the knowing agent. This is similar to Karl Popper's "epistemology without the knowing subject," though Popper would eschew the ascription of agency to articles (see Popper 1972, Chapters 3 and 4). Rather, on Popper's view the articles are part of the so-called Third World. These proposals are quite different from the one I am discussing here, and space does not permit me to discuss their merits and weaknesses.

[7] More recently, Beatty and Moore (2010) argue that it is dangerous for scientists to give the public the false impression that there is consensus in science when there is not. Beatty and Moore claim that the public would have more trust in science if they were made aware of the disagreements amongst scientists. Attempting to give the impression of a consensus when there is not one inevitably makes the public suspicious. Beatty and Moore claim that "the authority of a scientific body is not undermined by questioning, but rather depends on it" (2010, 211). This is an empirical question and deserves further study.

with large collaborative projects credit is diffused or dissipated. In a recent study of postdoctoral researchers working in the life sciences, Ruth Müller (2012) found that young scientists experience a tremendous amount of anxiety over authorship. The postdoctoral researchers Müller interviewed believe that they will be unable to have an academic career in science if they do not publish articles where they are listed as first author. Indeed, the postdocs were concerned that as more authors are listed on an article the value of the article is diminished, even when they are listed as the first author. These are not new concerns. Already in 1995, in a poll of members of the large research teams in high-energy physics, "the European Committee for Future Accelerators . . . found some seventy-five percent of their respondents disliked present publication habits, largely because it damaged the possibility of career advancement and of receiving credit for their work" (Galison 2003, 341). Consider the situation a scientist faces when she is listed as one of 1,681 authors on a paper (see Newman 2001, 406). This is an extreme case, though a real one. But even scientists involved in smaller research teams may feel alienated from their work.

Third, there is some evidence that there is a diffusion of both credit and responsibility in research produced by teams of scientists. When things go wrong, for example, when questions are raised about the integrity of data, often no individual scientist takes responsibility. As a result, the group as a collective or corporate body is left responsible.

Consider the practices involved in the retraction of scientific publications that have been found to be problematic. The current practices illustrate how responsibility for research produced collaboratively is evaded. It is not uncommon for a research team publishing a retraction to explicitly note that a specific member of the research team "objects to the retraction of the paper" or "declined to be an author of the retraction" (see Böhlenius et al. 2007, 367; but also Puneet et al. 2013 and Buck 2010). Not surprisingly, it is sometimes the lead author of the article being retracted who is most reluctant to sign the published retraction. Some retractions are "editorial retractions" authored by the editor of the journal in which the questionable article was published, rather than by the researchers who authored the questionable article. These may include a statement specifying how many of the authors of the original paper "have agreed to retract their paper" (see, for example, Kennedy 2006, 335; see also Alberts 2011). This brief examination of the practices surrounding the retraction of published articles illustrates how challenging it is to get a team of researchers, that is, a *whole* team, to take responsibility for the integrity of the research they produce collaboratively. These behaviors make sense when we distinguish between an individual scientist's view and the view of the research team.

Part of the rationale for claiming that the view of a research team is irreducibly a collective view is that alternative explanations fail to account for the phenomena outlined above. The most obvious alternative explanation is a traditional individualist explanation, where every belief or view held is the view of some individual. The individualist thus offers a summative account of ascriptions of beliefs to groups. According to the individualist account, a group can be said to hold a view when most individual members of the group hold that view. But it is unclear how an individualist account can make sense of the phenomena I have identified above. Specifically, it is not clear how the individualist can account for the fact that sometimes individual scientists work with groups that profess to hold views (indeed they *publish* views) that are at odds with the views of the individual members of the group.

Fagan (2011) proposes an alternative to both the collectivist explanation I am defending here and the traditional individualist explanation. Fagan proposes what she calls an interactive view of collective knowledge. Here is how Fagan describes the formation of an interactive view.

> A group's interactive belief may correspond to the belief had by the greatest number of members, the intersection of all members' beliefs, the belief had by all or most authoritative members, the average or median belief of members, the belief had by representatives of all *g*'s members [where *g* is a group] after deliberation in accordance with norms most members of *g* endorse. (Fagan 2011, 251)

Fagan's point is that there are many different ways a group can settle on an interactive view. She takes this to be a shortcoming of Gilbert's influential plural subject or collectivist account.

But there are problems with Fagan's view. Most importantly, it is not clear how the interactive view differs from the collectivist view, the view I am referring to as the plural subject account. Neither Gilbert nor the other philosophers who build on her plural subject account put constraints on how groups settle on their beliefs or views. Hence, the mechanisms identified by Fagan are all possible mechanisms for determining the view of the group. Thus, her so-called interactive view does not seem to be distinct from the plural subject or collectivist view defended here. Indeed, we could think of Fagan's analysis as providing an elaboration of the plural subject or collectivist view rather than challenging it.

As a matter of fact, different research teams are organized in different ways, and have different means to determine the content of the publications issuing from the group. In a recent piece in *Science*, Adrian Cho describes the differences between two research teams that work at the European Center for Nuclear Research (CERN). Both research teams involve over a thousand scientists and engineers. One group, ATLAS, has a democratic structure, with "a very well-framed constitution and rules" (Cho 2011, 1566). When it comes to determining the contents of their publications the ATLAS group "strives ... for consensus" (Cho 2011, 1566). The other group, CMS, has a "top down" management structure (1566).[8]

Both research teams are capable of adopting irreducibly collective views, even though the means by which each group settles on the view of the group are very different. Whereas the ATLAS group may involve more deliberation in an effort to reach a consensus on what should stand as the group's view, the views of the CMS group will be determined, to a larger extent, by those managing the group, the group

[8] Peter Galison provides detailed analyses of a number of different research institutions' authorship protocols, including protocols used at the Stanford Linear Accelerator Center, Fermilab, one of the Large Hadron Collaboration detectors at CERN, and two research teams at the Large Electron Positron, also at CERN (see Galison 2003, Section IV). These are all large teams of researchers, so they face similar challenges. But each research team has developed its own culture and practices.

leaders. In either case, though, the view settled on that stands as the group's view will be irreducibly the view of the group.

A determined individualist might raise the following objection. She might focus on my appeal to the distinction between belief and acceptance, and argue that when a research team settles on a view, the view is *accepted* by the individual members of the team, not the research team. In this way, there is no need to ascribe irreducibly collective views to groups. The only believing and accepting agents are individual scientists.[9] If this objection is correct, then the believing/accepting distinction renders it unnecessary to attribute views to groups. An individual scientist can accept a view as she works as a part of a research team even though she does not personally believe it.

This strategy for salvaging the individualist account is misguided. Further, it rests on an ambiguity with respect to what an individual member of a research team accepts when the team lets a view stand as the group's view. Let me clarify the nature of the ambiguity, and thus address the individualist's objection. I have argued that the views of a research team are irreducibly the view of the group. And what the individual members of the research team accept is that the view will stand as the view of the group. But this is not the same thing as letting the view stand as one's own view. That is, an individual does not have to accept all that the research team accepts. One only needs to accept that the views expressed in the publications of the group are the views of the group, and act accordingly. Indeed, one may in fact accept what the group accepts. But that is not required, and it is certainly both logically possible and possible in practice for an individual scientist to work as part of a research team that accepts a view that she does not accept.

Most importantly, the content of the two views I am distinguishing between here are different. When a member of a research team individually accepts a view the team accepts, she might accept the following proposition: "chemical x has such-and-such effects on population y." In contrast, when a member of a research team accepts that a particular view will stand as the view of the group, she accepts the following proposition: "the research team accepts that chemical x has such-and-such effects on population y." There is no inconsistency in an individual researcher accepting the latter claim but not accepting the former claim.

In summary, appeals to the notion of collective acceptance help us better understand scientists' behavior when they work collaboratively. And such appeals explain phenomena that are inexplicable from the traditional individualist perspective. Further, contrary to what Fagan suggests, no competitor view seems capable of accounting for these phenomena.

6.2 Some Normative Issues: Authorship and the Need for a New Referee

So far I have focused on the descriptive task of advancing an account of the relationship between individual scientists who work as part of a research team and the views

[9] I thank Carole Lee for drawing my attention to this particular objection and the need to address it.

that are published by such teams. I now want to turn to some normative issues surrounding the epistemic cultures created by collaborative research. I argue that the current practices for dealing with authorship and refereeing are unsatisfactory, given the changes in the way scientific research is typically done now.

The notion of the scientific author is in flux, and different fields are dealing with this issue in different ways. In discussing this issue, my analyses will draw on some existing studies which have focused on high-energy physics and the biomedical sciences and have contrasted the ways in which these two different fields operate. My concern is not specifically with high-energy physics and the biomedical sciences, but rather with fields in which Big Science is the norm, as in many areas of high-energy physics, and in fields that have explicitly attempted to construct authorship criteria, and thus face the crisis of authorship in collaborative research head-on, as has happened in the biomedical sciences.

The norms surrounding authorship at institutions like CERN, for example, are quite different from the traditional norms that made it clear who is responsible for what, and thus who got credit for a particular scientific discovery. At CERN "papers reporting experimental results ... have all members of the collaboration listed on the first page(s) ... The names are in alphabetical order; no clues as to who originated the research or performed large chunks of it can be derived from the list" (Knorr Cetina 1999, 167; see Biagioli 2003 for a description of the authorship norms at Fermilab).

The enormous collaborations that characterize research in high-energy physics have given rise to a new culture of scientific authorship. The collective ownership of the research results is evident in the norms surrounding the way the research is presented to scientific audiences. Once a paper is written, various members of the research team may be asked to present the paper at conferences (see Knorr Cetina 1999, 168–169). One cannot presume that the person presenting the paper played a key role directing the efforts of the group. Rather, the presenter may well be just another member of the group involved in the research.

It seems that the notion of authorship in high-energy physics is undergoing another significant change. The December 21, 2012, issue of *Science* reports on the recent discovery of the Higgs boson, the result of two decades of research at the Large Hadron Collider at CERN (see Della Negra et al. 2012, 1560). The issue includes an article *authored* by The CMS Collaboration and an article *authored* by The Atlas Collaboration. There is no longer a pretense that a list of individual authors would be informative. Indeed, readers of the articles are informed that "the complete list of authors and affiliations is available as supplementary material in *Science* Online" (The CMS Collaboration 2012, 1569; see also The Atlas Collaboration 2012, 1576). Further, the corresponding authors are to be reached at cms-spokesperson@cern.ch and atlas.publications@cern.ch. Thus, the contact information is not tied to a "lead author" or principal investigator (PI).

This is not an unprecedented practice. Knorr Cetina (1999) notes that "the work for which Carlo Rubbia at CERN received the Noble Prize in 1984 is referenced as UA1 Collaboration, G. Arnison et al. (1983a,b)" (Knorr Cetina 1999, 167). Galison argues that strains on the notion of authorship in high-energy physics have changed as the field has changed "from the bubble chamber work of the postwar period to the huge colliding beam collaborations of the 1980s and 1990s" (Galison 2003, 328). In the earlier phase, research groups were identified with specific individual group leaders (2003, 329); now, though, research in high-energy physics is done by groups that are "a more federal association of parts" (2003, 331).

There is a problem lurking in the epistemic culture of Big Science, a problem that is exacerbated by the new conception of authorship that seems to have emerged in high-energy physics. On the one hand, there is a prima facie reason to think that groups would be in a better position to ground scientific knowledge. After all, more people are standing behind the results reported; at any rate, more people scrutinized the results. But, on the other hand, there are apt to be circumstances under which groups are less trustworthy than individual scientists. Collaborative teams are like contemporary business corporations in some respects. Determining who is responsible for a problem can be very challenging. And the "corporate body" is sometimes left to blame, even though it can in no way be harmed by accusations of research misconduct. In the old culture of little science, where most articles were authored by individuals, scientists were in a very different situation. If one was found to be engaging in deceptive behavior, the individual scientist's own career was threatened. There was no possibility of blaming others. Thus, for the individual scientist, the stakes were quite high. Again, I emphasize, this problem is not unique to high-energy physics. It seems to affect any area of Big Science.

The investment in the "physical plant" that supports collaborative research seems to further shield scientists from accountability when problems arise. The large research facilities in high-energy physics and other areas of Big Science are like the large financial institutions operating today. They are too big to fail. It is hard to imagine dismantling CERN, for example. Steve Fuller (1999) suggests that the scale of investment in Big Science research projects is so great that changing course even when a particular line of research is proving unfruitful proves to be almost unthinkable. A whole community, even a political unit as large as a state or province, comes to be financially dependent on the research facility supporting the line of research. Given this broader social context, the epistemic accountability of the scientists may inadvertently be compromised.

Things are quite different in the biomedical sciences. Leading journals, like the *Journal of the American Medical Association* (*JAMA*), have developed author criteria. In fact, the medical journals are not the only ones to have addressed the question of authorship in this way. The American Psychological Association (APA) also has authorship guidelines (see Endersby 1996, 384–385). Coauthorship is very common in psychology. In fact, between 1984 and 1994, about 80% of the articles published in the *Journal of Personality and Social Psychology* listed more than one author (see Endersby 1996, 380, Table 2). And, more than 10% of the articles had four or more authors.

Let us consider the author criteria of *JAMA* in detail. These criteria identify conditions that need to be fulfilled in order for one to count as an author of a paper published in the journal. The conditions are quite stringent. It is worth quoting *JAMA*'s policies at length:

> Each author should have participated sufficiently in the work to take public responsibility for appropriate portions of the content. One or more authors should take responsibility for the integrity of the work as a whole,

> from inception to published article. Per the guidelines of the International Committee of Medical Journal Editors (ICMJE), authorship credit should be based only on (1) substantial contributions to conception and design, or acquisition of data, or analysis and interpretation of data; and (2) drafting the article or revising it critically for important intellectual content; and (3) final approval of the version to be published. Conditions 1, 2, and 3 must all be met.... In addition, authors are required to identify their contributions to the manuscript. (*JAMA* 2012)

As explicitly acknowledged, these guidelines are based on the ICMJE's recommendations. In the most recent version of the ICMJE's recommendations, published in August 2013, a fourth condition is added. According to these guidelines, authorship should also be "based on ... agreement to be accountable for *all aspects* of the work in ensuring that questions related to the accuracy or integrity of *any part* of the work are appropriately investigated and resolved" (ICMJE 2013; emphasis added). Elaborating, the ICMJE claims that "in addition to being accountable for the parts of the work he or she has done, an author should be able to identify which co-authors are responsible for specific other parts of the work" (ICMJE 2013).

One of the concerns that these policies seem designed to address is "honorary" authorship, where scientists are wrongfully listed as authors even though they made no creative contribution to a particular article. We can get a sense of the scope of the problem of honorary authorship in the biomedical sciences when we note that there were "twenty researchers worldwide ... [who] published at least once every 11.3 days throughout the 1980s" (Rennie et al. 1997, 580). These twenty researchers are thus each credited with authoring or coauthoring over 300 papers in a decade. That is a prodigious level of research productivity, a level that seems almost impossible to reconcile with *JAMA*'s and the ICMJE's ideals of accountability. Indeed, it is doubtful that these researchers could even list the titles of the 300+ articles they authored in that decade. I am in no position to know what these twenty authors contributed to the various papers credited to their names. But there is a tension between that level of productivity and the sort of accountability that the ICMJE guidelines and the *JAMA* author criteria aim for.

In an effort to curb such practices, the ICMJE explicitly addresses the issue of who is *not* to count as an author. The ICMJE state that "participation solely in the acquisition of funding or the collection of data does not justify authorship. General supervision of the research group is also not sufficient for authorship" (ICMJE 1997 in Biagioli 2003, 262; see also ICMJE 2013). Nevertheless the ICMJE claims that nonauthors should be acknowledged and their contributions explicitly identified (ICMJE 2013).

These *JAMA* and ICMJE guidelines highlight some important differences between different epistemic cultures in science. The guidelines about who counts as an author, as well as who does not count as one, are obviously inappropriate for Big Science, where some people focus narrowly on data collection, and some are concerned narrowly with the design and operation of instruments and equipment. Further, there is no way scientists working on a project with 400 other scientists could possibly identify the contributions of each member of the group. Indeed, it is unlikely that such a scientist would even know all of the members in group.

Though the authorship policies in the biomedical sciences aim to address some of the problems that arise in collaborative research and aim to ensure that scientists can be held accountable if mistakes are found, they are insensitive to the way that collaborative research is done. The most complicated and contentious issue is the requirement that authors identify their contributions. Given the nature of collaborative research, it is unreasonable to expect scientists to be able to parcel out the various contributions in an article to specific individuals. One of the benefits of working collaboratively is that some capacities and knowledge are emergent; the individuals working alone would not have these capacities and knowledge. Consequently, strictly speaking, these emergent capacities cannot be attributed to any *one* scientist working on the project. It would be like trying to identify which one of an animal's vital organs is responsible for its health. Each organ contributes something, but the health of the animal is not merely the result of each organ functioning well. It is the coordinating functioning of the organs that explains the health of the animal. The biomedical science journals, as represented by the *JAMA* author requirements and the ICMJE recommendations, seem to be stuck in an individualist frame of mind. The most recent ICMJE recommendations explicitly acknowledge the possibility of large multiauthor groups. But the recommendations still insist that "all members of the group named as authors should meet all four criteria of authorship" (ICMJE 2013).[10]

Thus, there are problems in both high-energy physics and the biomedical sciences. Whereas the biomedical sciences are stuck in an individualist frame of mind, where the various contributions in an article can be assigned to specific individual members of the research team that produced it, in high-energy physics corporate authorship is recognized, but these "agents" are as elusive as business corporations when trouble arises. When problems are discovered, it can be very challenging to determine who is responsible.

Mario Biagioli (2003) rightly notes that part of the problem is that the notion of authorship in science fulfills a variety of functions. In his analysis, Biagioli emphasizes the credit and responsibility functions, discussed above (see also Rennie et al. 1997, 580). But authorship plays, or at least has traditionally played, an important role in the assessment of individual scientists when they apply for academic jobs or promotions and tenure (see Biagioli 2003, 266).

I now want to examine a second normative issue encountered in collaborative research, specifically how collaborative research affects refereeing. I argue that significant changes in refereeing practices are needed in order for the refereeing process to fulfill its function.

It is widely acknowledged that refereeing plays an important role in science. But it seems that the refereeing process has not kept pace with the changes that have resulted from the increase in collaborative research and the scale of collaborative research projects. I want to outline two concerns pertaining to the refereeing of research

[10] It is worth emphasizing that sometimes authors can work collaboratively and determine with some degree of confidence which parts of the project each contributed. Zamora-Bonilla (2014) discusses such cases. But I doubt this is the norm, and my own experiences as a collaborating author left me quite uncertain about which parts of the published articles were my contributions, and which contributions were my collaborators' contributions.

results produced by collaborative teams. I will refer to the concerns as the problem of insufficient numbers and the problem of inadequate expertise.

First, consider the problem of insufficient numbers. Assuming there are a fixed number of scientists working in a scientific specialty, it seems that as collaborative research teams get larger the pool of qualified referees gets smaller. Consider a research field where papers are sometimes authored by 50 or more scientists. Clearly, with 50 scientists authoring a research report we have 50 fewer scientists to engage as referees. My concern here is not just one of administration and management. My concern is epistemic. As research projects engage a greater proportion of the scientists working in a specialty, there are fewer and fewer competent scientists available to referee the resulting research.[11]

In one sense, this problem is an old problem. A comparable problem arises when scientists rely on unique or rare instruments to generate data. If few scientists have access to the instruments that generate data in a field, then there will be a very restricted pool of scientists qualified to evaluate the research that builds on these data. Some familiarity with the instruments that generate the data is a prerequisite for competent refereeing. This problem seems to have arisen already in Robert Boyle's day, with the research employing what was then a high-technology scientific instrument, the air pump. Shapin and Schaffer recount the various challenges that Boyle faced when he presented his findings to the scientific community. Boyle had to convince his audience that he was a legitimate authority on the data that he presented, data generated by an instrument to which few people had access (see Shapin and Schaffer 1985). Similarly, Galileo's efforts to have telescopes distributed by the Medici family to important people across Europe had a similar aim. Galileo needed other astronomers to accept his claims about the hitherto unobserved phenomena (see Biagioli 1993, 57).

We seem caught in a bind with refereeing research produced by large research teams. A referee must be knowledgeable about the research area in order to be an effective judge of a research report. But we also expect referees to be impartial.[12] And one way to secure impartiality is to ensure that there is some distance between the referee and the group producing the research report. As research teams get larger, it will be more difficult to find qualified referees, scientists who are apt to be both impartial but knowledgeable. There thus appears to be an insufficient number of knowledgeable but impartial referees. Again, this is not a new problem. But the

[11] I recognize that not all scientific specialties are of equal size. In a subfield in which 600 scientists work, journal editors will have an easier time finding two or three qualified referees than in a subfield in which 150 scientists work. The point is that as articles are authored by larger groups of scientists, there is a smaller pool from which to draw referees. In my experience as an academic editor for *PLOS ONE*, I need to invite between three and eighteen reviewers before I secure commitments from two referees for a manuscript.

[12] Lee et al. (2013) provide a comprehensive review of the literature on bias and peer review. Their working definition of impartiality in the context of peer evaluations is "the ability for any reviewer to interpret and apply evaluative criteria in the same way in the assessment of a submission" (Lee et al. 2013, 4). Importantly, they recognize the limitations of this definition, as sometimes editors deliberately solicit opinions from two scientists with very different backgrounds and areas of expertise.

problem seems to be exacerbated by the conditions under which large research teams produce their research.

Consider the second concern, the problem of inadequate expertise. Given the nature of the research produced by collaborative research teams, there are few, if any, *individual* scientists qualified to evaluate the research produced. One of the principal reasons scientists collaborate with other scientists is that they lack the expertise to pursue the research they want to pursue (see Thagard 1999, Chapter 11). By working with other scientists, individuals are able to tap into others' expertise, expertise that they personally lack. Consequently, it is doubtful that a single scientist will have the requisite expertise to evaluate the research reports produced by large research teams. After all, if the referee were to embark on the same research herself, she would inevitably need to draw on the expertise of a comparable team of researchers. It seems that we need to change our refereeing practices to reflect this situation.

A shift similar to the one that has occurred with respect to the production of scientific research may have to occur with respect to the evaluation of research. Perhaps the solution is to have papers issuing from large collaborative research teams refereed by *teams* of referees. I am not merely suggesting that a single paper should be sent to numerous *individual* scientists for refereeing. That is already standard practice. Rather, I am suggesting that we may need to constitute referee-teams that work together to collectively evaluate a single paper under review. The teams would consist of a number of scientists with expertise in a variety of areas, thus making them collectively capable of evaluating the paper under consideration. The model for such a practice already exists. The National Institutes of Health (NIH) in the United States, for example, reviews grant proposals by committee. Individual scientists not only evaluate the proposals. They convene with other referees to discuss the merits and weaknesses of each proposal. In contrast, the National Science Foundation (NSF), the other leading source of government funding for research in the United States, uses a different system (see Cole 1992, 84 and 257, Note 3). The NSF has individual scientists independently rank proposals and then aggregates the scores. NIH actually has study sections consisting of a panel of scientists who review a set of proposals collectively.[13]

It is worth clarifying and underscoring the differences between the two concerns I raise here about refereeing. First, large collaborations in science cut into the number of scientists available to referee research reports. The larger a research team is, the smaller the pool of referees there will be who can assess the manuscripts produced. Second, because collaborative research brings together the expertise of two or more scientists, as research teams get larger it becomes less likely that there is any individual scientist qualified to assess the manuscripts submitted by large teams. Research teams are, after all, often created out of necessity. No single scientist has the expertise to conduct the research on her own. Consequently, we should expect that no single scientist will have the expertise to evaluate such research reports.

The high-energy physics research community has devised other means for critically reviewing research before publication which may alleviate, to some extent, the concerns I raise here (see Biagioli 2003, 273). The research groups at CERN, for

[13] I thank David Resnik of the National Institute of Environmental Health Sciences (NIEHS) for clarifying the nature of the assessment process used by the NIH.

example, have numerous venues and meetings at which findings and draft papers are subjected to critical scrutiny. At ATLAS,

> during collaboration weeks ... full days of working group or detector group and subgroup meetings [are] held simultaneously ... followed by detector-related steering group and institute meetings ... and these in turn are followed by a full day of plenary meetings and concluded with a collaboration board meeting at the end. (Knorr Cetina 1999, 174–175)

These various meetings provide valuable opportunities for feedback that may in fact address some of the concerns I raised above with respect to refereeing collaborative research.[14]

Examining the ways in which collaborative research is affecting the refereeing process provides an opportunity to reflect on the role of the refereeing process in science. One might be tempted to think that referees are assigned the task of ensuring that the claims made in a manuscript under consideration for publication are epistemically justified, and the paper constitutes a contribution to scientific knowledge. That picture is a little too simplistic. Further, it is to ask for more from referees than they can reasonably be expected to deliver. The refereeing process is just one part of the culling process that manuscripts face. It is just one of the safeguards in determining what gets published. Once published, though, manuscripts and their contents are subjected to additional and ongoing scrutiny as other scientists determine the value of the results reported in published articles. Scientists interested in using the claims reported in published articles are especially well situated for the task of further scrutinizing the methods, findings, and interpretations in articles. Their own research careers are in jeopardy if they draw on and cite questionable research. Thus it is not likely that they will just unreflectively defer to the judgment of anonymous referees (see Wray 2007b; Hull 1988).

6.3 Concluding Remarks

My aim has been to highlight a variety of ways in which the epistemic cultures in science are changing because of the increasing role played by collaborative research teams. First, I defended the plural subject view of research teams. The plural subject account makes sense of the following facts: (i) sometimes the individual members of a research team do not hold the views reported in the articles the research team publishes; (ii) scientists report anxieties over not receiving credit for their contributions when they work collaboratively; and (iii) accountability is more difficult to enforce when research produced by a team of scientists is found to be flawed.

Second, I have highlighted some normative issues raised by collaborative research. I argued that we have yet to find a way to deal with the problems surrounding

[14] One of the referees astutely asked about the process by which the papers announcing the research on the Higgs boson were refereed for publication in *Science*. I do not have this information, though it is a crucial piece of information for advancing our understanding of how scientists are dealing with the new problems posed by Big Science.

scientific authorship in the age of collaborative research. The notion of the scientific author is under strain and recent attempts by leading medical journals and in the high-energy physics research community do not adequately address the concerns. Finally, I examined some of the strains that collaboration is putting on refereeing. The nature of the research produced collaboratively makes it especially challenging for other scientists to evaluate.

Despite the concerns I raise about collaborative research, it is worth reminding ourselves why it is so prevalent today. Knowledge of some aspects of the world is only accessible to scientists if they work collaboratively. The emergent capabilities of research teams afford access to knowledge that is otherwise inaccessible. But we still need to be mindful of the costs we incur when we pursue such knowledge.

Acknowledgments

I thank Lori Nash, Kristina Rolin, Margaret Gilbert, Carole Lee, and Susann Wagenknecht for constructive feedback on earlier drafts. I also thank my audience at the conference on Epistemic Groups and Collaborative Research in Science, at the Maison des sciences de l'Homme Lorraine, at the Université de Lorraine, in December 2012. I especially thank Thomas Boyer-Kassem and Cyrille Imbert, the conference organizers, for inviting me to present this paper at the conference. I also thank the Maison des sciences de l'Homme Lorraine for supporting my travel to and lodging in Nancy. I thank Evelyn Brister, Karen Frost-Arnold, Greg Frost-Arnold, and John Capps for providing valuable feedback when I presented the paper to the Western New York Philosophy of Science Circle, at the Rochester Institute of Technology in July 2013. Finally, I thank the editors and referees for their constructive feedback.

References

Alberts, B. 2011. "Retraction," *Science*, 334 (December 23), 1636.

Andersen, H. 2010. "Joint Acceptance and Scientific Change: A Case Study," *Episteme*, 7: 3, 248–265.

ATLAS Collaboration. 2012. "A Particle Consistent with the Higgs Boson Observed with the ATLAS Detector at the Large Hadron Collider," *Science*, 338 (December 21), 1576–1582.

Beatty, J. 2006. "Masking Disagreement among Experts," *Episteme: A Journal of Social Epistemology*, 3: 1–2, 52–67.

Beatty, J., and A. Moore. 2010. "Should We Aim for Consensus?," *Episteme*, 7: 3, 198–214.

Biagioli, M. 1993. *Galileo Courtier: The Practice of Science in the Culture of Absolutism*. Chicago: University of Chicago Press.

Biagioli, M. 2003. "Rights or Rewards: Changing Frameworks of Scientific Authorship," in M. Biagioli and P. Galison (eds.), *Scientific Authorship: Credit and Intellectual Property in Science*. New York: Routledge, pages 253–279.

Böhlenius, H., S. Eriksson, F. Parcy, and O. Nilsson. 2007. "Retraction," *Science*, 316 (April 20), 367.

Bouvier, A. 2004. "Individual Beliefs and Collective Beliefs in Science and Philosophy: The Plural Subject and the Polyphonic Subject Accounts," *Philosophy of the Social Sciences*, 34: 3, 382–407.

Bouvier, A. 2010. "Passive Consensus and Active Commitment in the Sciences," *Episteme*, 7: 3, 185–197.

Buck, L. B. 2010. "Retraction," *Science*, 329 (September 24), 1598.
Cheon, H. 2014. "In What Sense Is Scientific Knowledge Collective Knowledge?," *Philosophy of the Social Sciences*, 44: 4, 407–423.
Cho, A. 2011. "Particle Physicists' New Extreme Teams: Life at the World's Biggest Atom Smasher Is an Odd Combination of Selfless Cooperation and Intense Competition," *Science*, 333 (September 16), 1564–1567.
CMS Collaboration. 2012. "A New Boson with a Mass of 125 GeV Observed with the CMS Experiment at the Large Hadron Collider," *Science*, 338 (December 21), 1569–1575.
Cohen, L. J. 1992. *An Essay on Belief and Acceptance*. Oxford: Clarendon Press.
Cole, S. 1992. *Making Science: Between Nature and Society*. Cambridge, MA: Harvard University Press.
de Ridder, J. 2014. "Epistemic Dependence and Collective Scientific Knowledge," *Synthese*, 191: 1, 37–53.
Della Negra, M., P. Jenni, and T. S. Virdee. 2012. "Journey in the Search for the Higgs Boson: The Atlas and CMS Experiments at the Large Hadron Collider," *Science*, 338 (December 21), 1560–1568.
Dragos, C. 2016. "Which Groups Have Scientific Knowledge? Wray vs. Rolin," *Social Epistemology*, 30: 5–6, 611–623. DOI: 10.1080/02691728.2016.1172361.
Durkheim, E. 1893/1984. *The Division of Labor in Society*, trans. by W. D. Halls, with an introduction by L. A. Coser. New York: Free Press.
Endersby, J. W. 1996. "Collaborative Research in the Social Sciences: Multiple Authorship and Publication Credit," *Social Science Quarterly*, 77: 2, 375–392.
Fagan, M. B. 2011. "Is There Collective Scientific Knowledge? Arguments from Explanation," *Philosophical Quarterly*, 61: 243, 247–269.
Fallis, D. 2007. "Collective Epistemic Goals," *Social Epistemology*, 21: 3, 267–280.
Fuller, S. 1999. *Governance of Science*. Buckingham: Open University Press.
Galison, P. 2003. "The Collective Author," in M. Biagioli and P. Galison (eds.), *Scientific Authorship: Credit and Intellectual Property in Science*. New York: Routledge, pages 325–355.
Giere, R. N. 2002. "Distributed Cognition in Epistemic Cultures," *Philosophy of Science*, 69: 637–644.
Gilbert, M. 2000. "Collective Belief and Scientific Change," in M. Gilbert (ed.), *Sociality and Responsibility: New Essays on Plural Subject Theory*. Lanham, MD: Rowman and Littlefield, pages 37–49.
Hull, D. L. 1988. *Science as a Process: An Evolutionary Account of the Social and Conceptual Development of Science*. Chicago: University of Chicago Press.
International Committee of Medical Journal Editors. 2013. "Recommendations for the Conduct, Reporting, Editing, and Publication of Scholarly Work in Medical Journals: Roles and Responsibilities of Authors, Contributors, Reviewers, Editors, Publishers, and Owners: Defining the Role of Authors and Contributors." http://www.icmje.org/roles_a.html (accessed September 18, 2013).
Journal of the American Medical Association. 2012. "*JAMA* Instructions for Authors." http://jama.jamanetwork.com/public/instructionsForAuthors.aspx (accessed August 12, 2012).
Kennedy, D. 2006. "Editorial Retraction," *Science*, 311 (January 20), 335.
Knorr Cetina, K. 1999. *Epistemic Cultures: How the Sciences Make Knowledge*. Cambridge, MA: Harvard University Press.
Lee, C. J., C. R. Sugimoto, G. Zhang, and B. Cronin. 2013. "Bias in Peer Review," *Journal of the American Society for Information Science and Technology*, 64: 1, 2–17.
Mathiesen, K. 2007. "Introduction to Special Issue of *Social Epistemology* on 'Collective Knowledge and Collective Knowers,'" *Social Epistemology*, 21: 3, 209–216.
Müller, R. 2012. "Collaborating in Life Science Research Groups: The Question of Authorship," *Higher Education Policy*, 25: 289–311.
Newman, M. E. J. 2001. "The Structure of Scientific Collaboration Networks," *Proceedings of the National Academy of Sciences*, 98: 2, 404–409.
Popper, K. R. 1972. *Objective Knowledge: An Evolutionary Approach*, revised edition. Oxford: Oxford University Press.

Price, D. J. de Solla. 1963. *Little Science, Big Science*. New York: Columbia University Press.
Puneet, P., C. T. Yap, L. Wong, L. Yulin, D. R. Koh, S. Moochhala, J. Pfeilschifter, and A. Huwiler. 2013. "Retraction," *Science*, 341 (July 26), 342.
Rennie, D., V. Yank, and L. Emanuel. 1997. "When Authorship Fails: A Proposal to Make Contributors Accountable," *Journal of the America Medical Association*, 278: 7, 579–585.
Rolin, K. 2008. "Science as Collective Knowledge," *Cognitive Systems Research*, 9: 1–2, 115–124.
Rolin, K. 2010. "Group Justification in Science," *Episteme*, 7: 3, 215–231.
Schmitt, F. F. 1994. "The Justification of Group Beliefs," in F. F. Schmitt (ed.), *Socializing Epistemology: The Social Dimensions of Knowledge*. Lanham: Rowman and Littlefield, pages 257–287.
Shapin, S., and S. Schaffer. 1985. *Leviathan and the Air-Pump: Hobbes, Boyle, and the Experimental Life*. Princeton. NJ: Princeton University Press.
Staley, K. W. 2007. "Evidential Collaborations: Epistemic and Pragmatic Considerations in 'Group Belief,'" *Social Epistemology*, 21: 3, 321–335.
Staley, K. W. 2010. "Evidence and Justification in Groups with Conflicting Background Beliefs," *Episteme*, 7: 3, 232–247.
Thagard, P. 1999. *How Scientists Explain Disease*. Princeton, NJ: Princeton University Press.
Trognon, A., and M. Batt. Unpublished: 2012. "From Results on Social Psychology of Collective Induction in Experimental Groups to Hypotheses on Epistemology of Collective Induction in Scientist's Groups." Paper presented at "Epistemic Groups and Collaborative Research in Science," Nancy, France, December 19.
Wray, K. B. 2001. "Collective Belief and Acceptance," *Synthese*, 129: 3, 319–333.
Wray, K. B. 2006. "Scientific Authorship in the Age of Collaborative Research," *Studies in History and Philosophy of Science*, 37: 505–514.
Wray, K. B. 2007a. "Who Has Scientific Knowledge?," *Social Epistemology*, 21: 3, 337–347.
Wray, K. B. 2007b. "Evaluating Scientists: Examining the Effects of Sexism and Nepotism," in H. Kincaid, J. Dupré, and A. Wylie (eds.), *Value-Free Science? Ideals and Illusions*. Oxford: Oxford University Press, pages 87–106.
Zamora-Bonilla, J. 2014. "The Nature of Co-authorship: A Note on Recognition Sharing and Scientific Argumentation," *Synthese*, 191: 1, 97–108.

7

Power, Bargaining, and Collaboration

JUSTIN BRUNER AND CAILIN O'CONNOR

7.1 Introduction

Over the past century, collaborative work has become increasingly common in academia (West et al., 2013). In the sciences, it has continued to spread as a practice across disciplines and countries (Sonnenwald, 2007; Sooryamoorthy, 2009). In fact, in most scientific fields, the *majority* of published work is now co-authored (Morrison et al., 2003). The social sciences, likewise, have seen an increase in collaborative work, particularly in fields like economics and psychology (Morrison et al., 2003; Hudson, 1996; Larivière et al., 2006). Even in the humanities, collaboration, though not the norm, is increasingly practiced (Larivière et al., 2006).

There are a number of reasons why this model of academic production is so popular. In some fields, increasing specialization, and the high cost of, for example, expensive instruments, mean that collaboration is necessary to produce meaningful research (Morrison et al., 2003). Many authors have argued, furthermore, that collaborating provides myriad benefits to individual researchers even when not strictly necessary. For example, collaboration can lead to greater academic productivity, which benefits individual researchers in hiring and promotion decisions (Morrison et al., 2003; Landry et al., 1996; Ferber and Teiman, 1980; Lee and Bozeman, 2005).[1] Co-authored papers in some areas are more likely to be accepted to top journals and are more likely to be cited (Laband, 1987; Gordon, 1980; Beaver and Rosen, 1979). Furthermore, it has been argued that collaboration can be useful to academics in defraying risk. When numbers of publications are important, it may make more sense to invest smaller amounts of time in many papers than to invest heavily in one paper that might not be accepted to a good journal (Hudson, 1996). Lastly, collaboration is important to academics for social reasons. Collaborators create stronger social networks in their fields, and may be viewed by peers as insiders (McDowell and Smith, 1992).

[1] Some authors have argued that, in fact, collaboration does not increase total productivity in terms of papers produced by a community (McDowell and Smith, 1992; Hollis, 2001). There is evidence, however, that departments and hiring bodies over-count co-authored papers (McDowell and Smith, 1992). If so, then even if collaboration does not increase total production, there is still a benefit to individual researchers for collaborating.

Boyer-Kassem, Mayo-Wilson, & Weisberg. *Scientific Collaboration and Collective Knowledge*.

These clear cut advantages to collaborative work raise ethical questions, however.[2] The structure of academia is such that rewards for successful work, such as promotion, prizes, and salary increases, tend to accrue to individuals. Furthermore, author order, which is the main proxy used to determine credit for academic collaborations, is inherently unequal. Co-authors must occupy different places in the author ordering. This means that academics must decide who will be best positioned to receive these individual rewards. Further benefits to collaboration, such as opportunities to present at prestigious conferences, must likewise be divvied up. How are the fruits of collaboration divided in academic communities? Are these divisions equitable? Are there ways in which these divisions systematically disadvantage certain groups?

There are, at least, two types of potential disadvantage along these lines that academics might worry about. The first occurs between peers when one peer belongs to an underrepresented group in a field. For example, West et al. (2013) and Sugimoto (2013) have shown that in many fields women are proportionally less likely than men to hold coveted first and last author positions.[3] A second potential disadvantage might arise as a result of the hierarchical structure of academia (Goodyear et al., 1992; Keith-Spiegel and Koocher, 1985). It is common for professors to collaborate with graduate students or post-docs, tenured professors with non-tenured professors, and the like. In these situations, one member of a collaboration may have significantly more power than other members, and may take advantage of this position.

It has become increasingly clear that human behavior is deeply influenced not just by individual psychology, but by the networks of social interaction that humans engage in (Christakis and Fowler, 2009, 2007). It can be very difficult, however, to empirically observe causal patterns in the dynamic world of social engagement. In particular, for the phenomena we are interested in, dividing the fruits of collaborative labor, empirical evidence suggests that inequities arise, but does not necessarily explain the sources of these inequities. For this reason, we investigate possible causal pathways through which patterns of disadvantage can arise for certain groups in academic collaboration using formal models. We present two sets of evolutionary game theoretic models corresponding to the two types of possible disadvantage described above. The first set of models is from O'Connor and Bruner (2017). In this work, we examine disadvantages that underrepresented groups may undergo as a result of the dynamics of social interaction when a minority group learns to interact with a majority group. We also present a set of novel models looking at evolving bargaining scenarios where some actors are more powerful than other actors. We consider three possible types of power inequality and show that in each case the less powerful partners can end up at a bargaining disadvantage. As we will argue, in both sets of models inequities in

[2] For more on the ethics of collaboration, see Simon (1970); Fine and Kurdek (1993); Keith-Spiegel and Koocher (1985); Erlen et al. (1997); Zuckerman (1968).

[3] It may be that women in these fields are contributing less than men in collaborative research projects, and so no inequity is occurring. Likewise, one possible explanation for the over representation of men in last author positions could be the fact that older generations of academics simply have fewer women. It may be, however, that either male authors are more successfully negotiating for prestigious author positions, or journals are rejecting papers where women are first and last authors disproportionately.

bargaining over collaboration can arise even in populations of individuals who are not motivated by either implicit or explicit bias.

The chapter will proceed as follows. In section 2, we briefly discuss the formal framework we employ and its relevance for the phenomena under investigation. In section 3 we discuss the work of O'Connor and Bruner (2017) and related work. In section 4, we present original results on effects of power in evolving collaborative scenarios. We conclude by discussing the relevance of these results to real-world academic communities.

7.2 Models and Methods

Evolutionary game theory is a formal framework for modeling how the behaviors of groups of actors engaged in strategic interactions change over time. These methods were first developed to model change by natural selection, but were quickly adopted by scientists and social scientists to model change due to cultural learning in humans. Evolutionary game theory is a good framework for investigating the phenomena of interest here because collaboration, and bargaining over collaboration, are strategic scenarios—each player's outcome is influenced by the behavior of his or her interactive partners. Furthermore, behaviors in these scenarios are learned both from community members and through individual experience with past interactive partners. In other words, these behaviors change over time in systematic ways, making them an appropriate target for evolutionary analysis.

An evolutionary game theoretic model starts, unsurprisingly, with a *game*. A game is defined by a set of *players*, a set of *strategies*, or possible actions for each player, *information* each player has about the game, and *payoffs* to each player for every possible combination of strategies chosen. Evolutionary game theoretic models usually assume that a population of actors is playing a game, and then use *dynamics* to model how strategies in such a population will change over time. The particular dynamics we will employ here are the replicator dynamics. This model of change is extremely general. It assumes that the proportion of a population playing a certain strategy will change as a function of how successful that strategy is for those using it. In particular, a strategy that is more successful than average will proliferate in a population and one that is less successful will be outcompeted.[4] Notably, the replicator dynamics have been shown to bear deep formal similarities to dynamics for learning by imitation in groups (or cultural evolution) (Weibull, 1997), and to dynamics for individual learning (Hopkins, 2002; Börgers and Sarin, 1997). In this chapter, we will use the replicator dynamics to represent these types of change. We assume

[4] The discrete time version of the replicator dynamics will be employed here as we present simulation results. The continuous time version of the replicator dynamics can be used to obtain analytic results, but these are difficult to calculate in many cases and unnecessary for our investigation. The discrete time dynamics give updates of a population distribution at discrete steps or generations. They are formulated as $x_i' = x_i\left(\frac{f_i(x)}{\sum_{j=1}^{n} f_j(x)x_j)}\right)$ where x_i is the proportion of a population playing strategy i, $f_i(x)$ is the fitness of type i in the population state x, and $\sum_{j=1}^{n} f_j(x)x_j$ is the average population fitness in this state.

that academics, like other actors, learn to behave in their own best interest through individual experience and by mimicking successful peers.[5]

In order to build an appropriate model of populations of collaborating academics, we first must consider the strategic scenarios that occur when academics collaborate. The first thing to note is that collaboration is a joint endeavor where two actors decide to cooperate in order to reap a reward that is either greater than that for individual work, or is unobtainable as an individual. The second thing to note is that in the process of collaborative labor, the actors must decide who will be better positioned to reap the fruits of this labor. As discussed in the introduction, author position in academia is often used as a proxy for credit by hiring bodies, peers, and academic institutions, often in ways that significantly impact academics. A higher number of first authorships, for example, can, in some disciplines, lead to perks like promotion, prizes, tenure, and higher salary (Fine and Kurdek, 1993). Academics in collaboration must typically determine author order and thus choose which collaborative partners will be most likely to benefit most from the endeavor.

The game we employ as our basic model of collaboration in academia captures the strategic scenario just described. In the collaboration game, two actors first decide whether or not to cooperate, and then, if they choose to cooperate, decide how to divide the fruits of cooperation.[6] The collaboration game can be thought of as an amalgamation of two commonly studied games. Actors first play what is referred to as a stag hunt. The stag hunt is a model of cooperation under risk. It is motivated by the following scenario. Two hunters can either choose to hunt stag or hare. The payoff for successfully hunting hare is lower than the payoff for successfully hunting stag (less meat). But two hunters are needed to take down a stag, while one alone can catch a hare. If one actor hunts stag and her partner does not, she gets nothing. Figure 7.1 shows a *payoff table* for this game, which includes payoffs to each player for any possible combination of strategies.

This first part of the collaboration game models whether academics will collaborate. As pointed out, the payoff for doing solo research is often less than that for collaborative research, and so deciding to cooperate in academia can be thought of as hunting a stag, rather than hunting small game alone. There is some inherent risk to

Player 1 \ *Player 2*	**Stag**	**Hare**
Stag	5,5	0,3
Hare	3,0	3,3

Figure 7.1 A payoff table for the stag hunt. Rows list strategies for player 1. Columns list strategies for player 2. Payoff for player 1 is listed first and player 2 second.

[5] This is not to imply that these are the only processes by which academics learn or choose to behave, but research on human learning in games indicates that in general humans, like all animals, will repeat actions that benefit them (Roth and Erev, 1995).

[6] This game is discussed at length by Wagner (2012). It is equivalent to a Nash demand game where players have an outside option, which will be discussed in section 4.

collaborative effort, however. As many academics can attest to, if a co-author fails to perform his or her duties on a joint project, fellow collaborators are left in the lurch. On a more insidious note, a researcher who falsifies data or inadequately supports research claims may put her co-authors at risk of academic dishonesty (Strange, 2008; Frost-Arnold, 2013). These possibilities are captured in the game by the fact that a stag hunter with a hare hunting partner will receive a low payoff.

In the collaboration game, if both agents choose to hunt stag (cooperate), they then decide how to divide the resources gained through this cooperative endeavor (in this case, to decide author order) by playing a Nash demand game. The Nash demand game is a model of simultaneous bargaining.[7] Two actors want to divide a resource, and each may demand some portion of it (50%, for example, or 10% or 92%). If the demands do not exceed the total resource, both actors receive their demand. If the demands are too high, though, the actors receive payoffs called the disagreement point (often nothing). This corresponds to a scenario in which bargaining fails. Figure 7.2 shows a payoff table for a simplified Nash demand game. In this game actors can demand either 4, 5, or 6 of a total resource of 10 (we refer to these demands as Low, Med, and High). In the rest of the chapter, we will model bargaining with a restricted game of this sort.[8]

It is tempting to interpret these demands literally when thinking about bargaining in collaborative scenarios. High is first author, Low is second author. Given that levels of effort tend to differ in collaborative work, it is more fruitful here to think of these demands as relative to effort. The High demand can be thought of as a demand for a relatively prestigious author position compared to effort exerted on the paper. The Low demand can be thought of as agreeing to a relatively poor author position compared to effort. The Med demand can be thought of as an attempt to distribute credit equitably.

The full payoff table for one version of the collaboration game is shown in Figure 7.3. As is clear from the figure, the actors may hunt hare, or may hunt stag and make either a low, medium, or high bargaining demand of their opponent. It should be noted that in this game, as in the other two just described, the payoffs themselves are less important than the relationships between payoffs. What unites a set of different payoff

		Player 2		
		Low	**Med**	**High**
Player 1	**Low**	4,4	4,5	4,6
	Med	5,4	5,5	0,0
	High	6,4	0,0	0,0

Figure 7.2 A payoff table for the Nash demand game. Rows list strategies for player 1. Columns list strategies for player 2. Payoff for player 1 is listed first and player 2 second.

[7] This game is so-called because it was introduced by Nash. It is also sometimes called the "Nash bargaining game" or the "divide the dollar" game (Nash, 1950, 1953).

[8] Similar restrictions have been employed in previous evolutionary analyses of the Nash demand game (Skyrms, 1994, 1996; Skyrms and Zollman, 2010; Young, 1993; Binmore, 2008; Alexander and Skyrms, 1999; Alexander, 2000).

Player 2

Player 1		Stag-Low	Stag-Med	Stag-High	Hare
	Stag-Low	4,4	4,5	4,6	0,3
	Stag-Med	5,4	5,5	0,0	0,3
	Stag-High	6,4	0,0	0,0	0,3
	Hare	3,0	3,0	3,0	3,3

Figure 7.3 A payoff table for the collaboration game. Rows model strategies for player 1. Columns model strategies for player 2. Payoff for player 1 is listed first and player 2 second.

tables as one "game" is the strategic scenario these relationships induce. As such, in our analyses of these games, we will sometimes vary payoffs to investigate how these changes influence evolutionary outcomes.

We introduce the collaboration game as a good model of bargaining over collaboration. In some cases, though, the basic Nash demand game may be more appropriate. The collaboration game assumes that players can opt out of collaborative arrangements (by hunting hare), but in some areas of academia this will not be possible. In particular, in the lab sciences collaboration between professors and graduate students is essential for all involved.[9] Furthermore, increasing specialization means that collaboration between lab groups is sometimes necessary to complete a project. Throughout the chapter, we discuss evolutionary analyses of both sorts of games and the relevance of these analyses to bargaining in academia.

Before moving on, it will be useful to say something about the *Nash equilibria* of the games just described. A Nash equilibrium for a game is a set of player strategies where neither player can deviate and improve her payoff. These strategy sets are thought of as special because once players arrive at a Nash equilibrium, they will not be incentivized to change. In evolutionary analyses, Nash equilibria are important because populations tend to evolve to them.[10] The Nash demand game has three pure strategy Nash equilibria—Low vs. High, Med vs. Med, and High vs. Low.[11] At any of these outcomes, the entire resource is divided. If either player deviates she either demands less (and so receives less) or demands more and reaches the disagreement point. The version of the collaboration game shown here has four pure strategy Nash equilibria—the strategy sets where players take Stag-Low vs. Stag-High, Stag-Med vs. Stag-Med, Stag-High vs. Stag-Low and Hare vs. Hare. Altering the payoffs for this game, though, will alter the set of equilibria. If, for example, the Hare payoff is higher than the low bargaining demand, the equilibria will be Stag-Med vs. Stag-Med and Hare

[9] Graduate students can, of course, switch laboratories. The point is that collaboration itself is unavoidable.

[10] To be more specific, stable outcomes for the replicator dynamics are always Nash equilibria, but not all Nash equilibria are evolutionarily stable.

[11] Pure strategies are those where players always choose the same action. Alternatively, game theorists consider mixed strategies, where players probabilistically decide how to behave. We only discuss pure strategy equilibria here as these are more relevant to our evolutionary analyses.

vs. Hare. If the payoff for Hare is higher than that for the medium bargaining demand, only Hare vs. Hare will be an equilibrium. To summarize, these equilibria represent situations where one agent demands more, the other demands, more, the two agents make equal, or fair demands, of each other, or they do not cooperate.

We have now described the basic games employed throughout the chapter. In the next section, we will review previous work using these games to see how minority groups may be disadvantaged when bargaining over collaboration. In section 4 we will look at some variations of these games, which will be described, and justified, there, to address how power dynamics in academia can lead to similar inequities.

7.3 Minority Populations and Bargaining

O'Connor and Bruner (2017) investigate how underrepresented groups may be disadvantaged in academia as a result of the dynamics of social situations.[12] We employ models that represent two interacting populations where members from each population can condition their behavior on the population their interactive partner hails from (henceforth referred to as the "type" of the agent).[13] In academia, such a setup might represent a field in which both male and female researchers behave differently when interacting with men than with women. There is empirical evidence that real-world researchers are prone to such conditional strategic behavior. For example, there is often backlash or anger against women (but not men) who behave assertively in bargaining scenarios (Tinsley et al., 2009; Bowles et al., 2007). It has been shown that researchers, when presented with otherwise identical male and female academic job candidates, are more likely to hire the male and more likely to offer him a higher salary (Moss-Racusin et al., 2012; Steinpreis et al., 1999). Similar bias has been found to be directed at both black and gay job candidates (Tilcsik, 2011; Bertrand and Mullainathan, 2003). And the list goes on.[14]

In two-population models, like those we analyze in O'Connor and Bruner (2017), evolutionary outcomes can be significantly different than in one-population

[12] We do not imply there that this is the only way these groups are disadvantaged in academia, but rather explore how social dynamics can lead to disadvantage even when explicit barriers to entry against women and minorities have been removed from academic communities.

[13] This setup was also employed by Axtell et al. (2001) and Skyrms and Zollman (2010). Some of the results discussed in O'Connor and Bruner (2017) are due to Bruner (2017), who employs minority-majority population models like the ones discussed here.

[14] It should be noted that one assumption that goes into the models presented by O'Connor and Bruner (2017), and into the models we present in section 4, is that actors meet interactive partners randomly. This is obviously a simplifying assumptions since in many academic communities racial and gender groups cluster into subdisciplines. Our results should be thought of as applying to an idealized case where assortment is random (though, as we argue, they may also help explain *why* such clustering happens in the first place). Bruner (2017) presents similar models where actors are more likely to meet like types. He finds that this introduction of non-random assortment does not substantially effect the outcomes of his models.

models.[15] In other words, allowing agents to condition their behavior on type (male/female, white/black, gay/straight) can significantly influence how groups of people learn to behave in strategic scenarios. In part this is because in a game with this sort of conditioning, a player's strategy set changes. In the Nash demand game with types, strategies are no longer "Med" or "Low," but are now of the form "play Med with in-group members and play Low with out-group members." For simplicity sake, we will refer to these strategies using an ordered pair like <Med, Low> where the first place represents an in-group strategy and the second an out-group strategy. Significantly, with these new strategies, new equilibria where actors reach a different equilibrium with in- and out-group members are possible. Consider, for example, a situation where women play <Med, Low> and men play <Med, High>. Here both groups divide resources equitably with in-group members, but when men and women bargain, men take home a greater portion of the resource. Under our interpretation, this would correspond to a scenario where same-gender co-authors tend to negotiate equitable authorship agreements, but between genders men negotiate for more favorable authorship positions. This second type of equilibrium will play an important role in the results we discuss throughout the chapter.

Axtell et al. (2001) look at simulations of such two-population models of bargaining and find that this type of equilibrium, which they interpret as reflecting "discriminatory norms," commonly arises.[16] In particular, situations like the one just mentioned, where both groups are fair with their own type, but discriminate against the other, commonly occur in their evolving populations. It is worth noting that these outcomes cannot occur in single-population models in which conditional strategies are not employed. In single populations, fair outcomes, i.e., those where the entire population evolves to demand 5, are most common in evolutionary models (Skyrms, 1996, 1994; Ellingsen, 1997; Young, 1993).[17]

Bruner (2017) extends Axtell's results to show that under many conditions when one of two groups playing a Nash demand game is in the minority, it is significantly more likely that this group will end up disadvantaged when bargaining with majority types. (They demand Low of the majority, who demand High in response.) In general, the smaller the minority, the greater the chances that the group will be disadvantaged. The reason for this relies on the dynamics of social interaction when a small group interacts with a larger one. Neither minority nor majority types interact with minorities very often as a result of their small numbers. For this reason, minority types are strategically relatively unimportant. For minorities, on the other hand,

[15] This is also the case in two-population evolutionary models of bargaining scenarios that employ different dynamics as in Young (1993) and Binmore et al. (2003).

[16] These authors do not use the replicator dynamics, but instead consider actors who choose best responses to expected opponent behavior based on some limited set of memories about past opponent behavior. Similar results are found for the replicator dynamics (Bruner, 2017). A more detailed comparison of these two dynamics is outside the scope of this chapter.

[17] Sometimes a portion of the population will evolve to demand Low while another portion demands High, but in this case there is no particular discrimination (Skyrms, 1996). Those demanding High do so to everyone, even though they get nothing when paired with others who demand the bulk of the resource.

the majority group is very important. Minorities quickly learn to respond to what majorities are doing. In bargaining scenarios, this often means learning to make small demands, rather than risk reaching the disagreement point again and again. Once minority members learn to make small demands, majority types slowly learn that they can take advantage of this situation by demanding more.[18] O'Connor and Bruner observe that in these types of outcomes, minority members are doubly disadvantaged because they almost always meet majority types for interaction. This means that they are receiving lower payoffs both in cooperative and in bargaining scenarios during the majority of their interactions. Note that the smaller the population, the more likely the disadvantaged outcome and the more detrimental this outcome. This may be particularly bad for intersectional groups that are very small (such as, for example, black women in a field where people of color *and* women are underrepresented). Complicated dynamics may be occurring for such groups, though, making this is an area for further research.

O'Connor and Bruner (2017) investigate whether similar effects can occur in the collaboration game. We present simulations of majority-minority population models and find that minority populations are more likely than majority populations to end up at a bargaining disadvantage when they collaborate with out-group types. In particular, we find that when the payoff to playing hare (non-collaborative work) is low, collaboration is ubiquitous but minority types are much more likely to arrive at a bargaining disadvantage than majority types. These results, and those from Bruner, indicate that minority groups may be more likely to end up systematically disadvantaged when bargaining in academia, and in particular when bargaining over collaboration.

We also investigate whether these types of inequitable outcomes disincentivize collaboration for minority members. As noted, collaboration can greatly benefit academics, and so if underrepresented groups are disincentivized from collaboration, this could lead to negative career outcomes. Assuming that collaborative efforts will result in a bargaining disadvantage for minority types, we indeed find that minority types are less likely to engage in between-group collaboration in the first place (by instead choosing to hunt hare). This is increasingly likely the worse their bargaining disadvantage.[19] Furthermore, this effect is more pronounced when the minority population is smaller. Again, the disadvantage to minority populations at these equilibria is especially problematic because minority types tend to frequently meet majority types. This means that in these simulations minorities are very likely to either collaborate with majority types and end up disadvantaged, or else avoid such collaborations and take the lower hare-hunting payoff.

[18] See Bruner (2017) for a more detailed description of this effect. In some situations, Bruner found that minority populations actually were advantaged by the same effect. This could occur if, for example, it benefited minorities to quickly learn to demand more of majority types because of the particular details of the game they played.

[19] This was modeled by assuming that the demands available to players in the Nash demand game varied. Actors could always choose an equal split (5), but the unequal choices ranged from 5.1 and 4.9 to 7.8 and 2.2. The more unequal the split, the more likely it was that minority groups would avoid collaboration.

One thing to note about this latter set of models is that they can also be thought of as representing situations where factors outside both collaborators' control ensure that the advantages that accrue to one type are greater than to those of another. Consider, for example, a case where a white academic and an academic of color collaborate. Now suppose that those evaluating this work assume that the academic of color contributed less than their collaborator (irregardless of author ordering) and, as a result, provide fewer benefits to them (promotions, tenure, salary, prizes, etc.). The results just cited indicate that in cases like this the disadvantaged types will be less likely to engage in collaboration since they are less likely to obtain the usual benefits from such an arrangement.

One might point out that if minority types in an academic community are disadvantaged when collaborating with majority types, they can just collaborate with other minorities. For small minority groups, though, it will be unrealistic, and potentially harmful, to restrict collaborative endeavors to like types. McDowell and Smith (1992), for example, argue that observed productivity gaps between male and female economists might, in part, be due to gender sorting in collaboration. They argue that if females tend to co-author with females only, they lose chances to collaborate, especially in departments with few women.

We will now move on to the second set of models on disadvantage in bargaining over collaboration. In the conclusion of the chapter we will further discuss the implications of the models just described, their fit with real-world phenomena, and their explanatory power.

7.4 Power and Bargaining

We present a novel investigation of another type of situation in which certain academic groups might be disadvantaged in collaborative work. Academic communities are arranged in an explicit hierarchy. Full professors outrank associate professors, who outrank assistant professors. Professors outrank post-docs who outrank graduate students. Even within these categories, there are unofficial hierarchies where famous or highly successful academics outrank the less famous, professors at research universities outrank those at teaching schools, community colleges, etc., and tenure track professors outrank those in non-tenure track positions.[20] Beyond hierarchical considerations, it is also worth noting that academics from wealthier schools and nations, or with more substantial grants may have unequal relationships with academics from poorer schools and nations and without impressive funding.

It is no surprise that power can lead to bargaining advantages (as we will discuss in more detail below), and authors have discussed how such power imbalances can lead to unethical behavior in academia particularly (Strange, 2008). In many academic disciplines, explicit norms of authorship have been proposed and adopted to help

[20] We do not mean to imply that these hierarchies totally order all academics. For example, the hierarchical relationship between a promising assistant professor at a prestigious school and a less successful associate professor at a less prestigious school may be ambiguous. What matters here is that in many cases these hierarchical relationships do exist and are completely clear to both parties.

mitigate this problem (Eggert, 2011; Fine and Kurdek, 1993; Spiegel and Keith-Spiegel, 1970; Erlen et al., 1997). For example, in many STEM fields, it is assumed that PIs (principal investigators) will be the last authors listed on a paper. First author will be reserved for the graduate student or post-doc who did most of the work, and the remaining collaborators will be arrayed in between. This convention is intended to ensure that a graduate student is given proper credit for joint work with a professor. Such conventions do not, however, prevent all types of potential injustice in the division of credit. Even in the lab sciences, there are scenarios where norms of convention (about co-authorship) are silent, such as when two PIs equally collaborate on a paper, or two graduate students share the majority of work. Outside STEM fields, conventions of authorship are often less established, and it is unclear how and whether rank plays a part in determining who will take what author role.

There is evidence that even when clear conventions governing author order exist, there is flexibility in how said order is determined. For example, Nobel laureates have been found to be less likely to be first author on papers after receiving the prize, presumably because they feel comfortable giving this privileged position to a more needy colleague (Zuckerman, 1968). Similarly, senior faculty are less likely to be first author when co-authoring with graduate students than junior faculty are (Costa and Gatz, 1992). The upshot of all this is that there is reason to explore how certain types of hierarchical privilege can lead to bargaining power in academic collaborations, even though explicit norms governing co-authorship have been established in some cases.

To assess how the dynamics of hierarchical interactions, such as those in academia, can result in such systematic inequalities, we return to our model of bargaining. We explore three ways of capturing how strategic bargaining evolves in scenarios with power imbalances between the interacting parties. Note that "power," as we use it here, does not necessarily refer to a unified concept. Instead, we think there are a cluster of ways that certain academic positions are more 'powerful' than others. We explore how some of these can translate into an improved bargaining position.

7.4.1 Different disagreement points

Recall the notion of a disagreement point. This is the payoff that both players receive when bargaining breaks down in the Nash demand game. We have been assuming until now that both parties in our bargaining interactions have the same disagreement point (and that it is 0). This assumption can of course be relaxed, the idea being that if a compromise cannot be reached, one party is left in a better position than her counterpart. This could be the case for a number of reasons in an academic scenario. For example, publishing yet another paper is of little consequence to an established professor (tenure is achieved, reputation is established). For an unestablished researcher, on the other hand, a single paper might make the difference on the job market, or the tenure chopping block. Intuitively, one would expect that when two agents do not have the same disagreement point the individual with the higher one has a bargaining advantage. In other words, the fact that she fares better than her counterpart when bargaining breaks down can be leveraged to ensure she

Player 2

Player 1		Low	Med	High
	Low	4,4	4,5	4,6
	Med	5,4	5,5	D,d
	High	6,4	D,d	D,d

Figure 7.4 Payoff tables for a Nash demand game with different disagreement points. Rows model strategies for player 1. Columns model strategies for player 2. Payoff for player 1 is listed first and player 2 second.

receives a majority of the contested resource.[21] This intuition is borne out in both the theoretical and empirical literature. The famous Nash bargaining solution, for example, stipulates that, all else being equal, an increase in one agent's disagreement point simultaneously increases the amount she should expect when a compromise is made. Likewise, recent work in experimental economics has shown that the agent with the higher disagreement point is at a slight advantage in bargaining (Binmore et al., 1989, 1991).

Consider the game presented in Figure 7.4. This is nearly identical to the Nash demand game of Figure 7.2 with the important difference being that Player 1 and Player 2 no longer have the same disagreement point. We assume that $D > d$. In other words, Player 1 (the established agent) does better than Player 2 (the unestablished agent) when their demands are incompatible.

Assume that $d < 4$, i.e., that the unestablished researcher always benefits from collaboration. If $D > 4$ then the situation in which Player 1 demands Low and Player 2 demands High is no longer an equilibrium. Similarly, if $D > 5$ then the sole pure-strategy equilibrium of this game is for Player 1 to demand the High and for Player 2 to concede. So one tentative conclusion we can draw from this is that for certain disagreement points the established agent should only be willing to collaborate if she receives the bulk of the benefits produced from said collaboration (a good author position relative to effort). We now go on to detail how such systematic inequalities can naturally emerge for less extreme values of D.

We consider a two-population model where the two groups are of equal size.[22] We once again assume that the strategies employed by established and unestablished types change as determined by the replicator dynamics. Note that in these simulations types switch *strategies*, but not *roles*. In other words, a professor cannot become a post-doc, but can change what action she takes when interacting with post-docs. Figure 7.5 details our results. In particular we track the proportion of simulations that result in one of three strategy pairings between populations (High vs. Low, Med vs. Med, and Low vs. High) as the disagreement point of the established population increases. When $d = D = 0$ it is just as likely that a norm favoring the

[21] To clarify, if bargaining breaks down the joint project is not completed and both agents are stuck with the unequal status quo.

[22] All of the simulation results presented in this chapter were conducted in MatLab 2012. For each set of parameter values we ran 1,000 simulations for 5,000 generations.

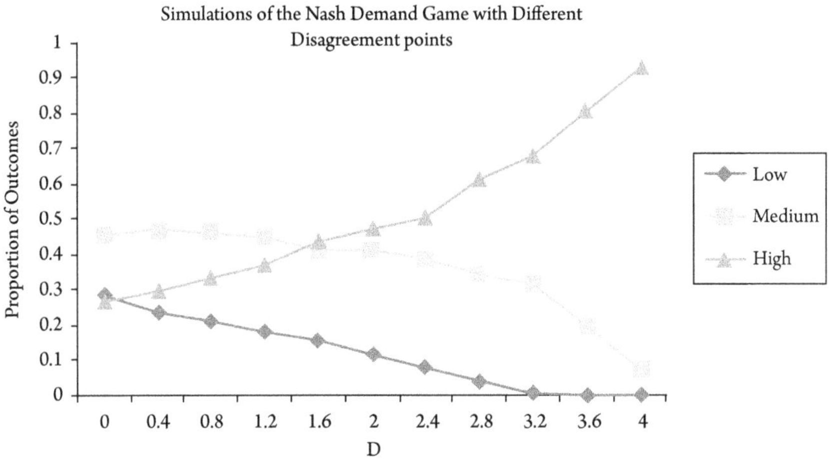

Figure 7.5 Simulation results for the game presented in Figure 7.4. The y-axis shows the proportion of simulations that result in senior academics demanding Low, Medium, and High as D increases (x-axis) and $d = 0$.

established population emerges as it is for a norm favoring the unestablished. Yet as the disagreement point of the established population increases, the proportion of simulations that head to the equilibrium in which the established population demands High increases. Not surprisingly, as D nears 5 it becomes overwhelmingly likely that the established population will demand the high amount when interacting with the unestablished population. The number of equal splits plummets at this point because achieving the equal split no longer significantly benefits the established player. These results indicate that if established academics suffer less than unestablished research partners when collaborative efforts break down, they may learn to demand and receive better author positions.

It is worth noting that a different interpretation is available for the levels of the disagreement points in these models. Nash (1953) argued that the value of disagreement points could be established not only through exogenous forces, but through threats made during pre-play discussion in bargaining scenarios. In other words, one actor could threaten to mete out punishment to the other and thus lower the other's payoff should bargaining fail. These "threatpoints" fell out of favor because other theorists challenged the ability of most bargainers to make plausible threats of this sort. In cases where one actor is clearly more powerful than the other, however, it is plausible that credible threats could be made. Academia is such a case. An established academic need not even explicitly threaten graduate students, post-docs, or early career colleagues with some sort of punishment (poor reviews, failure to mentor, refusal to write letters of recommendation, etc.) in the case of bargaining breakdown in order for such an implicit threat to exist. A graduate student collaborating with a powerful academic may be extremely conscious of the fact that should she anger this powerful colleague things will likely go poorly for her. Such threats can induce a payoff table like the one in Figure 7.4.

Player 2

Player 1		Low	Med	High
	Low	4+B, 4+b	4+B, 5+b	4+B, 6+b
	Med	5+B, 4+b	5+B, 5+b	0,0
	High	6+B, 4+b	B,b	0,0

Figure 7.6 Payoff tables for a Nash demand game with different background payoffs. Rows model strategies for Player 1. Columns model strategies for Player 2. Payoff for Player 1 is listed first and Player 2 second.

7.4.2 Different background payoffs

Another way of modeling the hierarchical structure of academia is to assume that established academics have multiple projects they are working on simultaneously, while those new to the profession are banking on successfully completing a small handful of papers (some of which may be co-authored). What this means is that when two individuals come to negotiate the terms of a collaborative endeavor, the more senior academic will have a higher "background payoff." Furthermore, for the less experienced academic, collaboration with his more senior colleague could constitute a large chunk of his early career research. Figure 7.6 presents a game which captures this scenario. B and b refer to the background payoff of established and unestablished researchers, respectively. In other words, these are the payoffs both agents get regardless of whether or not collaboration is successful. We assume that any benefits derived from collaboration will be in addition to these background payoffs. Finally, we make the reasonable assumption that $B > b$.

One may wonder whether this is just a different way of representing the game from 4.1 (Figure 7.4). This is not the case. In the game just presented (Figure 7.6) both parties stand to gain equally (in absolute terms) from successful collaboration. In the game from Figure 7.4, the junior researcher benefits more from collaboration than the senior researcher. Furthermore, for this new game, the senior researcher stands to gain from collaboration for all parameter values. Recall that this was not the case in Section 4.1. If $D > 6$, the senior researcher would actually receive lower payoffs for any collaborative outcome than for a failure to collaborate. For this reason, one might expect that this new scenario would be less likely to yield unequal outcomes, as both researchers always care about collaborating and all collaboration outcomes are Nash equilibria for all parameter values.[23]

Turning to the replicator dynamics, we can again track how likely a norm favoring senior academics is to emerge. Figure 7.7 presents simulation results. It is clear that the magnitude of the background payoff significantly affects outcomes. The larger the difference in background payoffs, the more likely the established type is

[23] This is the case because the addition of B and b to the payoffs in Figure 7.6 will always maintain the ordering of each player's payoffs. If 5 is larger than 4, $5 + B$ is larger than $4 + B$. This means that player preferences across actions will remain the same, and so the pure strategy Nash equilibria cannot change.

to end up demanding High.[24] The explanation for this is similar to that provided in Bruner (2017) and O'Connor and Bruner (2017). As described, we found that in majority-minority bargaining scenarios the larger group often ends up demanding High of the minority. This is due, as discussed, to an asymmetry in the learning environments of the two groups. One population (the minority population) is much more sensitive to strategic behavior of the other population (the majority group). This means the minority quickly adapts to the behavior of the majority while the majority is by and large insensitive to the minority population's behavior (see the previous section for more on this). A learning asymmetry is also present when the two populations have different background payoffs. Recall that strategies which do better than the population average grow in size under the replicator dynamics. Furthermore, strategies that do significantly better than the average grow rapidly, while those that only do slightly better grow at a slower rate. Thus background payoffs play a large role in determining the speed at which change occurs. If the background payoff is small then slight payoff differences between strategies are magnified and the population undergoes change rather quickly. Alternatively, when the background payoff is large, small payoff differences between strategies result in only minimal change from generation to generation. Established academics are slower to adapt, which in turn bestows upon them a bargaining advantage. These results indicate that if certain collaborative efforts are relatively unimportant for established academics, as compared to unestablished academics, systematic bargaining disadvantages may develop for early career researchers.[25]

7.4.3 Different outside options

We consider one final way in which an established academic may have bargaining power over an unestablished collaborative partner. Suppose that either party may choose to opt out of collaborating before bargaining begins and choose to engage in individual work instead. Formally, such a scenario is identical to a collaboration game (where players can choose either to bargain or to hunt hare). It may be, in such a case, that one player's outside option (hare payoff) is significantly better than the other's, and that this inequity can lead to unequal bargaining positions. If one player has a good enough outside option, cooperative scenarios where she does not receive a significant portion

[24] The proportion of simulations that result in the senior academics demanding High continues to increase as their background payoff, B, increases. Eventually, however, this proportion of simulations levels off at around 60. For $B = 200$, 57.6 percent of simulations result in established agents demanding High, and this percentage increases to 58.5 percent when $B = 500$.

[25] In these simulations, as discussed, we use the discrete time replicator dynamics. Under the continuous time version of these dynamics, the effect just described cannot occur as a result of adding a constant to the payoffs for one population of actors. This is due to differences in the formulations of these dynamics. However, as mentioned above, the results we see are due to differences in learning speed between the two groups. If one reformulates the continuous time replicator dynamics so that one population learns faster than the other, very similar results are seen. Relatedly, see Bergstrom and Lachmann (2003), who show that similar effects occur in biology when two species are interacting and one evolves at a slower rate than the other (this is the so-called Red King effect).

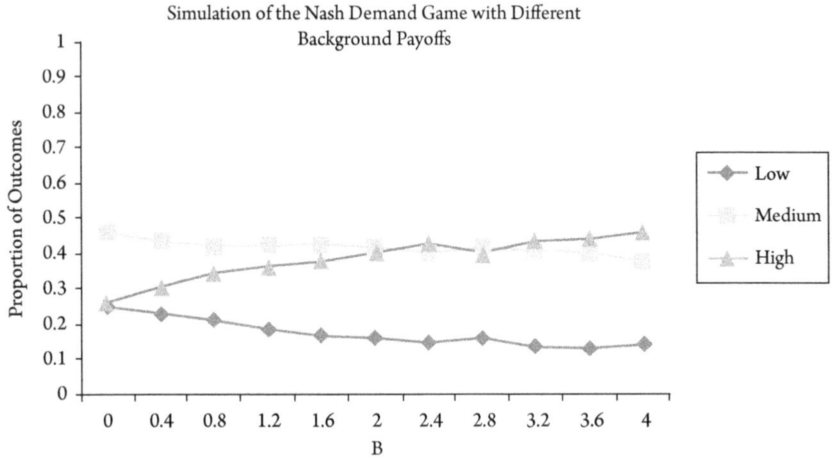

Figure 7.7 Simulation results for the game presented in Figure 7.6. The y-axis represents the proportion of simulations that result in senior academics demanding Low, Medium, and High as *B* (the base payoff of the more senior academic) increases.

Player 2

Player 1		Stag-Low	Stag-Med	Stag-High	Hare
	Stag-Low	4,4	4,5	4,6	0,h
	Stag-Med	5,4	5,5	0,0	0,h
	Stag-High	6,4	0,0	0,0	0,h
	Hare	H,0	H,0	H,0	H,h

Figure 7.8 Payoff table for the collaboration game with unequal payoffs for Hare. Rows model strategies for Player 1. Columns model strategies for Player 2. Payoff for Player 1 is listed first and Player 2 second.

of the resource are no longer Nash equilibria, meaning that she will be disincentivized from accepting a low bargaining demand (or, in our interpretation, a poor author position).[26] It is likely, for example, that an established academic can expect more significant payoffs from individual work—acceptance to respected journals, citations, invitations to present, etc.—than an unestablished partner. Alternatively, it could be the case that when a more experienced researcher collaborates with an unestablished academic the senior researcher must exert effort to mentor her counterpart. This is an extra cost that the established agent takes on, and thus the net benefit of collaboration is smaller for her. It is also an added benefit for the unestablished researcher, making her net benefit to collaboration higher. We model these scenarios with a collaboration game where the payoff to hunting hare is greater for the established academic. The payoff table for this game is shown in Figure 7.8.

[26] For an analysis of equilibrium selection in these types of games see Harsanyi and Selten (1988).

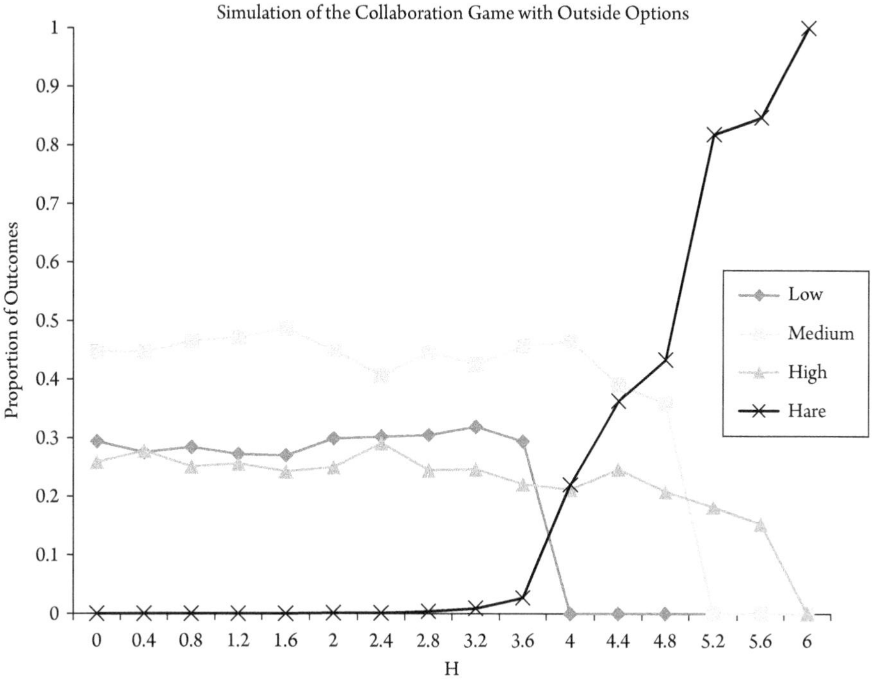

Figure 7.9 Simulation results for the game presented in Figure 7.8. The y-axis represents the proportion of simulations that result in senior academics demanding Low, Medium, High, and Hare as *H* increases (for h = 0).

Note that the game here includes differences in what agents get when they decide to work alone rather than to collaborate at all, but if they do collaborate, and collaboration fails, they both get nothing. In the game in Figure 7.4, on the other hand, there is no outside option. Researchers *must* collaborate, but if it goes south, one player ends up in a better position.[27] This said, there are some similarities between the games.

In the bargaining game with unequal disagreement points (Figure 7.4), if the disagreement point for the established researcher is high, she will be disincentivized to collaborate unless she receives a significant portion of the resources generated. In this scenario, likewise, if $H > 4$ the outcome where the established researcher receives the low bargaining demand will no longer be an equilibrium. If $H > 5$, the only collaborative outcome that remains an equilibrium is the one where the established researcher demands 6. (Of course, if $H > 6$, there will be no collaborative equilibria.) Unsurprisingly, results of our evolutionary simulations are consistent with this (see Figure 7.9). As *H* increases, outcomes where the established academics demand Low disappear, followed by Med, and then High. At the same time, the probability that the

[27] There is reason to think these two conditions will often show up together. (A more established researcher will do well both when bargaining collapses and when opting out of collaboration.) We keep these assumptions in separate models here to better understand their individual effects on collaboration.

established academic chooses to go it alone instead of collaborating increases and is inevitable once $H > 6$.[28]

These last results indicate that if senior academics reap greater benefits than junior colleagues from individual labor, they may learn to collaborate only in cases where they are well compensated for their efforts, i.e., only cases where they receive prestigious author positions.

7.5 Conclusion

We will start with separate discussions of the two sets of models presented, and what they tell us about the target phenomena. We will then discuss the explanatory force of the type of formal methods we use here to explore bargaining disadvantage in collaboration.

The results from O'Connor and Bruner (2017) indicate that social dynamical forces may lead to disadvantage for minority groups in collaboration and in bargaining over collaboration. It is worth noting that the results cited need not be the effect of any sort of pernicious behavior or intent on the parts of the involved parties. They arise on three conditions: 1) that individuals condition behavior based on type of interactive partner, 2) that one type of interactive partner is in the minority, and 3) that individuals evolve their behavior in ways that benefit themselves. Arguably the only morally questionable acts among these involve conditioning behavior on the type of partner, though even this is not universally thought to be morally wrong. Of course, one could also argue that individuals always have a moral obligation to protect colleagues from their own self-promotion, but the point is that actors in these models need not be trying to disadvantage underrepresented groups or even be aware that they are.

The set of models on power in bargaining over collaboration, likewise, indicates that social dynamical effects can lead to systematic patterns of bargaining that disadvantage unestablished academics such as graduate students, post-docs, and early career researchers. These effects should not be terribly surprising given that, as mentioned, previous research has indicated that in non-evolutionary scenarios, i.e., one-time interactions where rational actors make decisions, powerful positions can

[28] Perhaps surprisingly, when both outside options are low, even if one is significantly higher than the other, the population with the lower outside option is not disadvantaged in the subsequent bargaining. As is seen in Figure 7.9, the established academic is just as likely to demand the majority of the resource as her unestablished peer while H is low. This can be explained in the following way. In this game, if players decide to collaborate, the structure is the same as a Nash demand game with identical disagreement points. In other words, if the hare payoff does not deter collaboration, it becomes irrelevant to the bargaining interaction. To give a clearer understanding of the evolutionary process occurring, the evolution of hare hunters can be thought of as separate, in a way, from the evolution of the other players. The hare hunters do not bargain with other players and so influence the payoffs of other types equally. (I.e., a hare hunter will equally disadvantage the Low, Med, and High players when she meets them.) Hare players will either disappear if the other players outperform them, or take over the population if they outperform the rest. Whichever of these they do, they influence the other three types equally and so do not significantly alter the evolutionary process happening *between* these other types.

lead to bargaining advantages. Once again, though, it is significant to note that in our results, actors need not actively decide to pursue their own interest at the expense of an interacting partner. Instead, processes of learning through experience and from peers can lead to these inequitable outcomes. For these sorts of processes to occur there are now only two necessary preconditions 1) actors condition strategies on interactive partners and 2) actors learn to behave in their own best interest. In this case, arguably, neither of these behaviors is ethically problematic. It is impossible to avoid conditioning behavior on hierarchical positions in academia. Indeed, this sort of conditioning is expected in almost every sort of interaction across the explicit hierarchies of academia. And, obviously, learning to do self-beneficial things is as natural to humans, and indeed almost every type of animal, as breathing. Once again, the effects we describe can be seen when actors are not behaving in ways that would usually be considered objectionable, and when actors may not even be aware of the systematic disadvantages occurring in their fields. Of course, this is not to imply that the resulting behaviors in both sets of models are unproblematic. Rather our models show that from a starting point without regular, recognizable bias, populations can arrive at behaviors consistent with bias and discrimination simply by slowly learning to benefit themselves.

The results of both sets of models fit with certain empirical observations. In regards to underrepresented groups, Wilkie and Allen (1975) find that women sociologists tend to be credited as equal collaborators when co-authoring with women (and men with men) but that in male-female collaborations men are more likely to be credited as the principal contributors. Furthermore, there are indications that women may be disincentivized from collaborating by disadvantages of these sorts (West et al., 2013; McDowell and Smith, 1992; Ferber and Teiman, 1980; Boschini and Sjögren, 2007) and that when they do collaborate they are more likely to choose women partners (McDowell and Smith, 1992; Ferber and Teiman, 1980). Del Carmen and Bing (2000) show that black criminologists are significantly less likely to collaborate than white ones. Furthermore, it has been observed that underrepresented groups sometimes tend to cluster in academic subdisciplines (Botts et al., 2014).

Empirical work also shows that ethical concerns over co-authorship with senior colleagues do arise in academic communities (Goodyear et al., 1992; Costa and Gatz, 1992; Frost-Arnold, 2013; Bennett and Taylor, 2003; Eggert, 2011). In some areas, there has been increasing concern over senior faculty who pressure junior faculty for undeserved authorship positions (Bennett and Taylor, 2003; Kwok, 2005). Such practices have been argued to damage junior faculty who then receive less credit from hiring and promotion bodies (Frost-Arnold, 2013; Strange, 2008).

The fit of empirical data with certain results from the presented models is suggestive. But it is possible that, despite this fit, the causal processes occurring in these models do not actually reflect those happening in real epistemic communities. For this reason, it seems appropriate to ask: what is the value of these models? Do they provide genuine explanatory insight? Obviously these evolutionary game theoretic models are extremely simple, and so can only be thought of as idealized representations of complex processes such as those occurring in the social sphere. In these particular models, for example, actors choose strategies concurrently. In reality, complete breakdowns of bargaining are presumably rare because actors can engage in an ongoing discussion of the situation at hand.[29] Despite simplification, however, these sorts of models provide insight into causal processes in these very complex situations that is difficult to obtain otherwise. As just discussed, they allow us, for example, to list preconditions for these effects to occur that would not have been obvious without these models.

There are several important things to be said about the results just described which can mitigate worries about the explanatory power of these models. First, these results certainly provide evidence that such social dynamical effects are *possible*. At very least, they ought to give reason for academics to be on the lookout for the occurrence of these sorts of bargaining disadvantages even in groups of unbiased peers. Second, both sets of results are very robust. In other words, they are not relics of a particular model, but correspond to deep mathematical regularities in the sorts of setups we examine. Two-population outcomes arise for all sorts of games, in all sorts of parameter settings. We have already explained why. Two-population models have entire sets of new Nash equilibria that become evolutionarily relevant. As long as groups of interacting partners are conditioning behavior on types, these effects can and will arise. The minority effects observed in O'Connor and Bruner (2017) are furthermore extremely robust. Robust results, like these, are usually thought to be more likely to provide actual (instead of just possible) explanations of real-world phenomena, as the chances that the phenomena will fall under the set of conditions under which these outcomes occur are higher.

Third, the aspects of these mathematical models that lead to disadvantaged outcomes are easy to understand, and easy to verify empirically in the real world. It is clear that actors condition behavior on interactive partners (as established earlier in the chapter). It is clear that real-world actors learn strategic behavior from peers and from past experience. And it is the case that in many academic fields women and minorities are indeed underrepresented, and that established academics have certain types of power when compared to unestablished academics. These observations increase confidence that the effects seen in these models could actually be occurring in academic communities.

Now it seems appropriate to ask: Given these results, what should be done? Are there policy recommendations that these results support? Perhaps the most important takeaway from our work is one we have already emphasized. As we have seen, norms which disadvantage the vulnerable are likely to naturally emerge in the absence of formal or explicit rules regarding how credit should be allocated among co-authors. For this reason, we follow authors such as Erlen et al. (1997), Fine and Kurdek (1993), and Strange (2008) in recommending explicit communal policies be adopted governing co-authorship. Such policies will hopefully decrease chances of inequity in determining authorship. This, as argued, will not be a universal solution, as there will always be ambiguous or unusual cases where it is not clear how the collaboration fits with discipline standards. But we believe such measures are better than nothing.

[29] Another simplification that bears mentioning is the assumption that academics learn, from peers and through experience, to do whatever benefits them. Human behaviors are, of course, influenced by many complex factors, including overt ethical obligations, and our models do not account for these.

Importantly, work by West et al. (2013) indicates that some norms are better than others. These authors find that women are (or have been) underrepresented in first and last author positions in a number of sciences. In mathematics, however, where authorship is alphabetic, this effect is not observed.[30]

More generally we suggest that our results should help provide fodder for what ought to be an ongoing discussion in academia—how to ensure fairness in collaborative work. If academics are aware of the ways that inequities can arise as a result of the dynamics of social interaction, either between minority and majority groups or between groups in hierarchies, they will hopefully shape discussions of policy matters related to collaboration appropriately.

References

Alexander, J McKenzie (2000). "Evolutionary explanations of distributive justice." *Philosophy of Science*, *67*(3), 490–516.

Alexander, Jason and Brian Skyrms (1999). "Bargaining with neighbors: Is justice contagious?" *Journal of Philosophy*, *96*(11), 588–598.

Axtell, Robert L, Joshua M Epstein, and H Peyton Young (2001). "The emergence of classes in a multiagent bargaining model." *Social Dynamics*, 191–211.

Beaver, D deB and Richard Rosen (1979). "Studies in scientific collaboration Part III. Professionalization and the natural history of modern scientific co-authorship." *Scientometrics*, *1*(3), 231–245.

Bennett, Dianne M and David McD Taylor (2003). "Unethical practices in authorship of scientific papers." *Emergency Medicine*, *15*(3), 263–270.

Bergstrom, Carl T and Michael Lachmann (2003). "The Red King effect: When the slowest runner wins the coevolutionary race." *Proceedings of the National Academy of Sciences*, *100*(2), 593–598.

Bertrand, Marianne and Sendhil Mullainathan (2003). "Are Emily and Greg more employable than Lakisha and Jamal? A field experiment on labor market discrimination." *Technical report, National Bureau of Economic Research.*

Binmore, Ken (2008). "Do conventions need to be common knowledge?" *Topoi*, *27*, 17–27.

Binmore, Ken, Peter Morgan, Avner Snaked, and John Sutton (1991). "Do people exploit their bargaining power? An experimental study." *Games and Economic Behavior*, *3*(3), 295–322.

Binmore, Ken, Larry Samuelson, and Peyton Young (2003). "Equilibrium selection in bargaining models." *Games and Economic Behavior*, *45*(2), 296–328.

Binmore, Ken, Avner Shaked, and John Sutton (1989). "An outside option experiment." *The Quarterly Journal of Economics*, *104*(4), 753–770.

Börgers, Tilman and Rajiv Sarin (1997). "Learning through reinforcement and replicator dynamics." *Journal of Economic Theory*, *77*(1), 1–14.

Boschini, Anne and Anna Sjögren (2007). "Is team formation gender neutral? Evidence from coauthorship patterns." *Journal of Labor Economics*, *25*(2), 325–365.

Botts, Tina Fernandes, Liam Kofi Bright, Myisha Cherry, Guntur Mallarangeng, and Quayshawn Spencer (2014). "What is the state of blacks in philosophy?" *Critical Philosophy of Race*, *2*(2), 224–242.

Bowles, Hannah Riley, Linda Babcock, and Lei Lai (2007). "Social incentives for gender differences in the propensity to initiate negotiations: Sometimes it does hurt to ask." *Organizational Behavior and Human Decision Processes*, *103*(1), 84–103.

[30] Other conventions such as, for example, randomly determining authorship would similarly prevent such inequities.

Bruner, Justin (2017). "Minority (dis)advantage in population games." *Synthese*. DOI:10:1007/s11229-017-1487-8.

Christakis, Nicholas A and James H Fowler (2007). "The spread of obesity in a large social network over 32 years." *New England Journal of Medicine*, *357*(4), 370–379.

Christakis, Nicholas A and James H Fowler (2009). *Connected: The Surprising Power of Our Social Networks and How They Shape Our Lives*. Hachette Digital, Inc.

Costa, M Martin and Margaret Gatz (1992). "Determination of authorship credit in published dissertations." *Psychological Science*, *3*(6), 354–357.

Del Carmen, Alejandro and Robert L Bing (2000). "Academic productivity of African Americans in criminology and criminal justice." *Journal of Criminal Justice Education*, *11*(2), 237–249.

Eggert, Lucas D (2011). "Best practices for allocating appropriate credit and responsibility to authors of multi-authored articles." *Frontiers in Psychology*, *2*(196), 1–6.

Ellingsen, Tore (1997). "The evolution of bargaining behavior." *The Quarterly Journal of Economics*, *119*(2), 581–602.

Erlen, Judith A, Laura A Siminoff, Susan M Sereika, and Laura B Sutton (1997). "Multiple authorship: Issues and recommendations." *Journal of Professional Nursing*, *13*(4), 262–270.

Ferber, Marianne A and Michelle Teiman (1980). "Are women economists at a disadvantage in publishing journal articles?" *Eastern Economic Journal*, *6*(3/4), 189–193.

Fine, Mark A and Lawrence A Kurdek (1993). "Reflections on determining authorship credit and authorship order on faculty-student collaborations." *American Psychologist*, *48*(11), 1141.

Frost-Arnold, Karen (2013). "Moral trust & scientific collaboration." *Studies in History and Philosophy of Science Part A*, *44*(3), 301–310.

Goodyear, Rodney K, Clyde A Crego, and Michael W Johnston (1992). "Ethical issues in the supervision of student research: A study of critical incidents." *Professional Psychology: Research and Practice*, *23*(3), 203.

Gordon, Michael D (1980). "A critical reassessment of inferred relations between multiple authorship, scientific collaboration, the production of papers and their acceptance for publication." *Scientometrics*, *2*(3), 193–201.

Harsanyi, John C and Reinhard Selten (1988). *A General Theory of Equilibrium Selection in Games*. Cambridge, MA: MIT Press.

Hollis, Aidan (2001). "Co-authorship and the output of academic economists." *Labour Economics*, *8*(4), 503–530.

Hopkins, Ed (2002). "Two competing models of how people learn in games." *Econometrica*, *70*(6), 2141–2166.

Hudson, John (1996). "Trends in multi-authored papers in economics." *Journal of Economic Perspectives*, *10*(3), 153–158.

Keith-Spiegel, Patricia and Gerald P Koocher (1985). *Ethics in Psychology: Professional Standards and Cases*. New York: Crown Publishing Group/Random House.

Kwok, LS (2005). "The White Bull effect: Abusive coauthorship and publication parasitism." *Journal of Medical Ethics*, *31*(9), 554–556.

Laband, David N (1987). "A qualitative test of journal discrimination against women." *Eastern Economic Journal*, *13*(2), 149–153.

Landry, Réjean, Namatie Traore, and Benoît Godin (1996). "An econometric analysis of the effect of collaboration on academic research productivity." *Higher Education*, *32*(3), 283–301.

Larivi, Vincent, Yves Gingras, and Éric Archambault (2006). "Canadian collaboration networks: A comparative analysis of the natural sciences, social sciences and the humanities." *Scientometrics*, *68*(3), 519–533.

Lee, Sooho and Barry Bozeman (2005). "The impact of research collaboration on scientific productivity." *Social Studies of Science*, *35*(5), 673–702.

McDowell, John M and Janet Kiholm Smith (1992). "The effect of gender-sorting on propensity to coauthor: Implications for academic promotion." *Economic Inquiry*, *30*(1), 68–82.

Morrison, Philip S, Gill Dobbie, and Fiona J McDonald (2003). "Research collaboration among university scientists." *Higher Education Research & Development*, *22*(3), 275–296.

Moss-Racusin, Corinne A, John F Dovidio, Victoria L Brescoll, Mark J Graham, and Jo Handelsman (2012). "Science faculty's subtle gender biases favor male students." *Proceedings of the National Academy of Sciences*, *109*(41), 16474–16479.
Nash, John (1950). "The bargaining problem." *Econometrica: Journal of the Econometric Society*, *18*(2), 155–162.
Nash, John (1953). "Two-person cooperative games." *Econometrica: Journal of the Econometric Society*, *21*(1), 128–140.
O'Connor, Cailin and Justin Bruner (2017). "Dynamics and diversity in epistemic communities." Working Paper.
Roth, Alvin E and Ido Erev (1995). "Learning in extensive-form games: Experimental data and simple dynamic models in the intermediate term." *Games and Economic Behavior*, *8*(1), 164–212.
Simon, Julian L (1970). "A plan to improve the attribution of scholarly articles." *The American Sociologist*, *5*(3), 265–267.
Skyrms, B. (1996). *Evolution of the Social Contract*. Cambridge: Cambridge University Press.
Skyrms, Brian (1994). "Sex and justice." *Journal of Philosophy*, *91*(6), 305–320.
Skyrms, Brian and Kevin JS Zollman (2010). "Evolutionary considerations in the framing of social norms." *Politics, Philosophy & Economics*, *9*(3), 265–273.
Sonnenwald, Diane H (2007). "Scientific collaboration." *Annual Review of Information Science and Technology*, *41*(1), 643–681.
Sooryamoorthy, Radhamany (2009). "Collaboration and publication: How collaborative are scientists in South Africa?" *Scientometrics*, *80*(2), 419–439.
Spiegel, Don and Patricia Keith-Spiegel (1970). "Assignment of publication credits: Ethics and practices of psychologists." *American Psychologist*, *25*(8), 738.
Steinpreis, Rhea E, Katie A Anders, and Dawn Ritzke (1999). "The impact of gender on the review of the curricula vitae of job applicants and tenure candidates: A national empirical study." *Sex Roles*, *41*(7–8), 509–528.
Strange, Kevin (2008). "Authorship: why not just toss a coin?" *American Journal of Physiology-Cell Physiology*, *295*(3), C567–C575.
Sugimoto, Cassidy R (2013). "Global gender disparities in science." *Nature*, *504*, 211–213.
Tilcsik, András (2011). "Pride and prejudice: Employment discrimination against openly gay men in the United States." *American Journal of Sociology*, *117*(2), 586–626.
Tinsley, Catherine H, Sandra I Cheldelin, Andrea Kupfer Schneider, and Emily T Amanatullah (2009). "Women at the bargaining table: Pitfalls and prospects." *Negotiation Journal*, *25*(2), 233–248.
Wagner, Elliott O (2012). "Evolving to divide the fruits of cooperation." *Philosophy of Science*, *79*(1), 81–94.
Weibull, Jörgen W (1997). *Evolutionary Game Theory*. Cambridge, MA: MIT Press.
West, Jevin D, Jennifer Jacquet, Molly M King, Shelley J Correll, and Carl T Bergstrom (2013). "The role of gender in scholarly authorship." *PloS One*, *8*(7), e66212.
Wilkie, Jane Riblett and Irving Lewis Allen (1975). "Women sociologists and co-authorship with men." *The American Sociologist*, 19–24.
Young, H Peyton (1993). "An evolutionary model of bargaining." *Journal of Economic Theory*, *59*(1), 145–168.
Zuckerman, Harriet A (1968). "Patterns of name ordering among authors of scientific papers: A study of social symbolism and its ambiguity." *American Journal of Sociology*, *74*(3), 276–291.

PART FOUR

FROM INDIVIDUAL TO COLLECTION OPINION

8

A Clustering-Based Approach to Collective Beliefs

DENIS BONNAY

How do collective beliefs in a group depend on the mental states of members of the group? The answer depends on the kind of group. In organized groups, in which members share a sense of group membership, specific attitudes come into play. Members may jointly commit to upholding certain beliefs as members of the group (Gilbert, 1989). Or they may aim at maintaining some consistency regarding the beliefs attributed to the group at various moments of its history (Pettit, 2003). What about unorganized groups, in which individual beliefs about the matter at hand are all there is? Simply equating what the group thinks to what a majority of people in the group think does not always make sense. One reason is Condorcet's paradox and the fact that majoritarian aggregation may take us from consistent individual beliefs to an inconsistent group belief. Another reason is that if individual opinions are very diverse, considering average beliefs does not seem very interesting anyway. What is at stake then is not just aggregating individual beliefs but also identifying coherent subgroups, to which aggregation may be meaningfully applied.

The present chapter aims at providing some conceptual and formal underpinnings for doxastic clustering, that is, attribution of collective beliefs in which clustering and aggregation of beliefs are intertwined. The first four sections are devoted to a conceptual vindication of doxastic clustering. In Section 8.1, I distinguish two theses regarding supervenience of collective beliefs on individual mental states. According to the stronger supervenience thesis, collective beliefs in a group supervene on corresponding individual beliefs of group members. I argue further in Section 8.2 that strong supervenience is a reasonable thesis for unorganized groups. When strong supervenience holds, impossibility results in judgment aggregation theory raise questions for the very existence of collective beliefs, because they show that coherent collective beliefs may simply fail to exist, as recalled in Section 8.3. Actual practice regarding the study of beliefs in heterogenous groups suggests that this problem is side-stepped by attributing beliefs at the level of subgroups. This is doxastic clustering, which is introduced in Section 8.4: what matters is identifying different ways of thinking about a set of issues and identifying clusters within the population that exhibit each of these particular ways. The last three sections and the

Boyer-Kassem, Mayo-Wilson, & Weisberg. *Scientific Collaboration and Collective Knowledge*.

appendix are devoted to a formal analysis of doxastic clustering. In Section 8.5, I introduce a framework adapted from judgment aggregation theory, in which so-called clustering methods replace aggregation functions. In this framework, axioms spelling out desirable properties of clustering methods are laid down (Section 8.6). The claim in Section 8.7 is then that these axioms characterize minimization of Hamming distance within clusters, which corresponds to majority voting. Proofs are given in the Appendix.

8.1 Group Belief and Supervenience

We attribute beliefs to groups, and there is nothing wrong with this. Some opinions do not belong to individuals or do not belong just to some individuals. They are collective opinions of a group. But individuals are the primary holders of beliefs. Beliefs are mental states we attribute to agents endowed with a mind. Groups do not *really* have minds of their own. Groups may have beliefs or goals only because they consist of agents who do. Groups do not exist independently of their members and, in particular, they do not have mental attributes independently of their members. This is the philosophical problem of collective belief: we rightly attribute beliefs to groups, but we are nonetheless committed to ontological individualism.

These two commitments can be made compatible by accepting some form of supervenience of collective beliefs upon individual mental states. Collective entities which are not endowed with a mind of their own have mental states because these mental states supervene on individual mental states. Two groups not sharing the same opinions necessarily comprise members having different attitudes. There is no secret soul of a group which would allow two groups whose members have exactly the same beliefs, desires and other attitudes to have different group opinions. This is not to say that group beliefs merely summarize individuals beliefs. The patterns of dependency between individual attitudes and collective beliefs may well be very complex. As argued by List and Pettit (2011), this complexity is a key ingredient in explaining why we attribute beliefs to groups and not just to individuals. When the pattern of dependencies is intricate, it may well not be possible in practice to short-circuit talk of collective beliefs and directly interact with the group on the basis of the individual beliefs of its members. Treating the group itself as a collective agent endowed with beliefs of its own is justified in so far as it often is the only feasible option.

Granting that collective beliefs supervene on individual features, what is it exactly that they supervene upon? The content of the supervenience thesis crucially depends on what we take the relevant individual features to be. Let us distinguish two supervenience theses along these lines:[1]

[1] The idea of weak supervenience is put forward by List & Pettit (2011). It proves fruitful to push the analysis in terms of supervenience a step further and connect types of supervenience and types of groups, as I shall do in the following sections.

Strong Supervenience Thesis. *Collective opinion in a group G regarding a particular issue I at a given time t supervenes on the doxastic states of the members of G regarding I at t.*

Weak Supervenience Thesis. *Collective opinion in a group G regarding a particular issue I at a given time t supervenes on the mental states of the members of G.*

The two theses differ with respect to their supervenience basis. According to the first, stronger thesis, the supervenience basis is rather meager. The relevant individual beliefs are only the ones which directly correspond to the purported collective belief. According to the second, weaker thesis, the supervenience basis is rather rich. Relevant individual features are not limited to those directly considered for attribution to the group. Individual beliefs other than the beliefs at stake at the collective level matter. For example, whether the group accepts ϕ may depend not only on whether its members accept the particular proposition ϕ but also on the kind of reasons they would accept as proofs of ϕ. Also, individual attitudes other than beliefs matter. Whether the group accepts that ϕ may depend not only on whether each of its members accepts that ϕ but also on whether they are willing to defer to some designated member's judgement on ϕ. Such willingness belongs to the wider supervenience basis acknowledged on account of the the weak thesis, but it does not belong to the narrow basis featured in the strong thesis. Finally, past attitudes, not just present ones, matter. Whether a group might rightly be said to believe that ϕ may depend not only on whether its members currently accepts that ϕ, but also on whether the group has made previous doxastic commitments implying that ϕ.

Competing accounts of collective belief to be found in the literature may be classified regarding to the supervenience basis they go for: the larger the basis, the more independent collective beliefs are from individual beliefs. Summative accounts, as epitomized by Quinton (1975), uphold the strong supervenience thesis. List and Pettit (2011) explicitly accept a version of weak supervenience, but, at least for purposive groups, they would reject strong supervenience. Groups are considered as collective agents because their members are willing to maintain some degree of consistency, which also typically involves relating to past beliefs of the group and its members (Pettit, 2003). Gilbert (1989) similarly rejects strong supervenience and accepts weak supervenience, but her supervenience basis is even more remote from the one for strong supervenience. Beliefs of group members regarding ϕ are largely irrelevant when it comes to whether the group believes that ϕ. According to Gilbert, group beliefs involve rather different kinds of attitudes in its members, in particular joint acceptance.

8.2 Organized Groups versus Unorganized Groups

One does not necessarily need to choose between these accounts. Rather it seems that the connection between collective and individual beliefs is bound to depend upon the kind of group under consideration. Let us say that a group is organized if there is at least some level of awareness of group membership. For organized groups, there is no reason to a priori restrict the supervenience basis. Group beliefs are the collective making of the members of the group, and this making typically involves much more than the meager set of beliefs acknowledged on account of strong supervenience. The group may be a group 'à la Pettit,' an enduring collective entity driven by certain goals, such as a political party or a trade union. In order to properly push the goals it defends in an open context, the group will insist on enforcing some degree of inter-temporal consistency and some degree of rationality; it will be bound by past decisions and by the weight of reasons. When it comes to deciding what counts as collective beliefs, members are willing to ponder their own beliefs with considerations pertaining to consistency and accountability. The group may be a group 'à la Gilbert,' a temporary association of individuals based on mutual dedication and some constancy, say a poetry reading group. In order to maintain its cohesion, the group will rely on joint commitments. Group members commit themselves to publicly endorsing some beliefs conditionally on other members making the same commitment.

Obviously, much more would need to be said to properly match kinds of organized groups and kinds of dependence between collective beliefs and individual states. But the main point should be clear enough. In organized groups, what counts as collective beliefs depends on a potentially wide range of individual attitudes regarding the group. Therefore, the weak supervenience thesis is well suited as a thesis about collective beliefs in organized groups, and the strong supervenience thesis is not.

What about unorganized groups? The strong supervenience thesis now looks more reasonable. Arguably, when members have little or no awareness of group membership, when they do not care about what may count as collective beliefs for the group under consideration, collective beliefs are much more directly connected to the corresponding individual beliefs. What the UAW thinks about Ford's pension buyout is a complex matter, which does not depend only on what UAW members currently and privately think about the issue. It also depends on UAW history, on its past stands regarding workers' benefits in the automotive industry, on its expectations regarding the long-term consequences of such a move for pension plans, and so on. By contrast, there is no room for such complexity when considering, say, what the French think about transgenic seeds. At least when it comes to this issue, the French do not count as an organized group. Arguably, what the French think about transgenic seeds depends upon what each French man thinks about the issue, and nothing more. Believers in a French *esprit de la nation* might prefer examples about opinions in groups displaying even less unity: what do iPad users age 20–30 think about the Android operating system? The Apple marketing team is interested in this kind of collective opinion, even though there is nothing more to it than what each individual iPad user in the relevant age group thinks about the issue.

8.3 Supervenience and Impossibility Results

Functional dependence between individual and collective opinions has been formally investigated in the theory of judgment aggregation. Famous results in the field directly bear upon the distinction between weak and strong supervenience. First of all, the impossibility theorem due to List & Pettit, which is the fundamental theorem of judgement aggregation, says that some set of constraints on how collective opinions may be derived are not jointly satisfiable. Let us recall the result and discuss it in the light of the supervenience issue:[2]

Theorem. *For sufficiently rich agendas, there is no aggregation function satisfying universal domain, collective rationality, anonymity and independence.*

The agenda is a set of formulas representing issues under consideration. Aggregation functions map profiles of individual judgment sets into collective judgments. Profiles are naturally taken to represent opinions of members in the group, and the output of the aggregation function the opinion of the group. Universal domain, collective rationality, anonymity, and independence are desirable constraints on the aggregation function. Universal domain requires that collective opinion always exists; collective rationality requires that collective opinion be coherent and complete; anonymity is treating all members on a par; and independence is deciding upon a particular proposition only on the basis of what group members think about that proposition.

According to the strong supervenience thesis, collective opinions supervene on current individual opinions on the matter under consideration. When strong supervenience holds, the way individual beliefs determine collective beliefs (the relation of functional dependence modeled by the aggregation functions) satisfies independence and anonymity. Indeed, on account of strong supervenience, if individuals beliefs regarding an issue are fixed, collective beliefs are hereby fixed as well. This implies independence, because collective belief on ϕ is not allowed to depend on individual beliefs on propositions different from ϕ. The case for anonymity is a bit more tricky. What does it mean to supervene on individual beliefs? A belief may be individuated just by its content, as the belief that ϕ, or the belief that $\neg\phi$. Or a belief may be individuated in a more fine-grained manner by its content and its holder, as John Doe believing that ϕ. On the former construal of facts about individual beliefs, strong supervenience requires anonymity, on the latter, it does not. Which reading is adequate for the strong supervenience thesis? Arguably, unorganized groups do not come with distinctions among their members, at least if group belief is merely to be representative of individual beliefs. But attribution of group beliefs may pursue different goals, one may be willing to attribute true beliefs to the group, and distinguish members according to their competence, or one may be interested in the dynamics of beliefs, and distinguish members according to how influential they are. However, in the present setting, not so much hangs on anonymity. Doing without anonymity only yields one kind of aggregation functions, dictatorships in which collective beliefs are identified with the belief of one distinguished individual in the group. At any rate, such kind of functional dependence is not a very tempting model of collective belief in unorganized groups.

Given the connection between strong supervenience and conditions on aggregation functions in the impossibility theorem, the upshot is that collective opinion does not always exist in groups for which strong supervenience holds. Now judgment aggregation theory has also produced possibility results. They take mostly two forms, depending on whether universal domain or independence is relaxed. First, one

[2] The impossibility theorem goes back to List & Pettit (2002) and may be found in its now classical forms in Dietrich & List (2007) and Dokow & Holzman (2010).

may relax universal domain to try to characterize the circumstances under which independence, collective rationality, and anonymity can jointly be satisfied. As an example, it is obvious that plain consensus makes aggregation possible. But consensus is not the only way. As an example, when the distribution of opinions in the group is suitably unidimensional, majoritarian aggregation is coherent (List, 2003).[3] In such favorable circumstances, it is thus safe to attribute collective beliefs to unorganized groups. Second, consistency maintenance can be built-in in the aggregation procedure. One idea is to abide by a given aggregation function exactly when this is compatible with prior decisions. Such path-dependent procedures enforce collective rationality and forsake independence (List, 2004). They mimic the behavior of organized groups that are striving for at least some degree of coherence in the opinions which may be attributed to them. In particular, these groups are bound by their past decisions, on pain of being perceived as constantly changing their minds. These cases are typical failures of strong supervenience. What counts as the opinion of the group is determined by attitudes such as beliefs about logical connections between judgments at stake, or the desire to attain consistency, or past beliefs.

Thanks to the theory of judgment aggregation, we thus know that some level of homogeneity within the group is necessary and sufficient to make collective beliefs and strong supervenience compatible. We also know that weak (instead of strong) supervenience makes collective beliefs possible in organized groups. But what about unorganized groups in unfavorable circumstances? It seems we do not know much, and this will be the topic of the next section.

8.4 Doxastic Clustering in Unorganized Groups

When it comes to unorganized groups with a low degree of (first-order and meta-) consensus, should we be content with the negative lesson we have learned and conclude there are no collective beliefs in those cases? Let us have a closer look at a real-life example and get some more intuitions. What do the French think about transgenic seeds? Indeed, there may be no such thing. Opinions can be collected through a survey on a variety of issues, so as to put figures on the percentage of French people who think that research on transgenic seeds should be allowed, or that research should be allowed if it is conducted by public institutions, or that transgenic seeds may be used by farmers if research has proved innocuity, and so on. Arguably, the question is highly multi-dimensional and opinions are very diverse. We may well end-up with a discursive paradox, where what the majority thinks about the issue is incoherent, although individual opinions are coherent.

Even if we get lucky and stay safe from a Condorcet situation, it is likely that what the majority thinks is actually not so interesting. Averaging over diverse and conflicting opinions will yield average results that do not adequately reflect any actual kind of attitude in the French population toward genetically modified organisms. What

[3] A profile of judgment sets is unidimensional if individuals can be so ordered that, for every proposition in the agenda, people accepting that proposition are an initial or final segment. The result can be tightened by allowing in some more profiles.

is interesting in order to understand what the French think about transgenic seeds is to identify what the broad kinds of attitudes represented in the French population are and who the groups sharing these attitudes are. Maybe there are 'Deep Greens' who are against any kind of artificial intervention on organisms, 'Scientists' who think that transgenic is the way of progress, 'Skeptics' who think that the main problem is about science being manipulated by private interests, and so on. Knowing what the French think is about identifying these types of attitudes, and knowing how widespread they are among French people. Maybe we hear the Deep Greens a lot, but they are only 5% of the population; maybe we do not hear skeptics; but more than half of French citizens are skeptics, and so on. At any rate, answering the question what the French think is not about a Procrustean aggregation of these diverse views. It is rather about distinguishing these views and spelling them out.

This process is in keeping with standard practice in opinion research, as pursued by sociologists or market research professionals. Straightforward summative aggregation of opinions provides a summary of data ("53% of the French think that transgenic seeds are dangerous"). It often is the kind of result that makes it to the newspapers. But it is by no means the main output of the research, which rather consists in the kind of clustering of individual opinions I have just described. Let us see in some more detail how opinion research typically goes. First, a survey is designed and data is collected for a sample that is to be representative of the population of interest. Once this has been done, statistical methods, such as principal component analysis, are used to identify what the main dimensions structuring the issues under scrutiny are. The third stage consists in applying clustering algorithms to group together respondents who have similar opinions. Aggregation comes at the fourth and final stage, within clusters. Clusters are described in terms of what their collective beliefs are and also who the people belonging to a particular cluster are. Going back to our example about transgenic seeds, a survey would first be conducted among a representative sample of the French population. By means of an analysis of the respondents' answers, researchers would reveal at the second stage that trust in science, trust in agrobiotech corporations, or value of nature conservation are the key dimensions that shape respondents' opinions. Clustering is performed at the third stage and results in identifying subgroups within the representative sample sharing similar attitudes along the key dimensions. Finally, the clusters can be characterized in terms of the opinions they hold. For example, 89% of respondents within the first cluster are found to agree or strongly agree that humans should refrain from intervening in ecosystems. Assuming other beliefs in the cluster cohere, such a cluster rightfully gets baptized as the group of 'Deep Greens.'

According to this picture, attribution of collective beliefs in unorganized and heterogenous groups works in a very different way from belief attribution to organized groups or to unorganized groups whose members have homogenous beliefs. In the latter cases, belief attribution operates at the level of the group. That group is a given, and its identification precedes belief attribution. In the former case, belief attribution operates at the level of subgroups. Subgroups are not given, they are discovered, and beliefs are constitutive of subgroup membership. Moreover, the role played by coherence is extremely different. When the doxastic group is given, inconsistency is a threat to the existence of group belief. When the doxastic groups are to be discovered, coherence is a guide to the identification of these groups.

Before moving to the formal analysis of doxastic clustering, let us take a second look at the concept of collective belief in unorganized groups that is targeted here. One might object that two different notions have been summoned: one corresponding to some everyday attributions of collective beliefs to unorganized groups; the other being an artifact of political, sociological, or market research. The notion from folk psychology stems from our everyday practice when we speak about 'what the leftists think,' 'the opinion of the aesthetes,' or 'those whose think that ....' We are attributing beliefs to unorganized groups that are defined by the kinds of beliefs they entertain. Accordingly, membership in the group is largely defined in relation with the typical beliefs in the group. Being a leftist is largely, if not entirely, defined by holding some specific social and political beliefs. Being an aesthete is largely, if not entirely, defined by endorsing some specific tastes and values. Second, there is a purportedly scientific practice in sociological or market research that consists in describing what a given group of people think by identifying and describing what coherent subgroups think. In a study on what the French think about transgenic seeds or, more generally, about issues at the interface of ecology and technology, researchers may be led to isolate the Deep Greens and characterize a coherent set of beliefs about ecology and technology that defines what it is to be a Deep Green. As before, beliefs are attributed to unorganized groups that are defined by the kinds of beliefs they entertain, and someone belongs to the group depending on the kind of belief she entertains. There is nothing more to being a Deep Green than sharing a large amount of the beliefs which are typical of the Deep Greens.

I do not take these two approaches to correspond to two different notions, even if the means that are used in identifying the relevant subgroups and characterizing their beliefs are very different. In the first case, belief attribution is based on implicit social knowledge about what other people think, which may be acquired in a lot of different ways. By contrast, the sociologist will be using some systematic methods to register individual beliefs in a sample of the population of interest, often by means of surveys and interviews. In the first case, doxastic subgroups do not necessarily partition a fixed population, their identification heavily depends on matters of salience and interest — as a case in point, the Deep Greens might go largely unnoticed even though they constitute a natural doxastic unit when it comes to ecological issues. By contrast, in the second case, the aim is generally to provide a systematic understanding of the range and distribution of opinions within a given population on some given issues. It is therefore crucial to provide a partition of the population, without leaving some people aside.

Despite these differences, the notion of collective belief at work does not seem to differ. The relationship between individual beliefs and collective beliefs is of the same kind: strong supervenience holds. Moreover, the relationship between beliefs and group membership is the same: the group is a doxastic group; it is defined by the coherence and the similarities between individual beliefs of group members. It is a group because individual beliefs of its members can be adequately summarized, regarding the matters of interest, in terms of a shared group belief. As a consequence, the question whether, in a given context, such doxastic groups exist seems to make sense only from an interpretive perspective. Doxastic groups do not exist in virtue of their members relating to the group as such, by contrast to what is happening in organized groups. They exist in so far as it is possible for us to interpret a given landscape of individual opinions as partitioned into coherent doxastic units.[4]

8.5 Formal Framework

In the remainder of this chapter, I would like to sketch a formal framework for doxastic clustering. Following social choice theory, the literature in judgment aggregation has taken the group as a given. The strive for consistency stems from this initial constraint. Consistency is hard to achieve because individual judgments to be aggregated may be arbitrarily diverse. And judgment aggregation theory is silent on group constitution. A formal account of doxastic clustering in unorganized groups has to go a different way. The task is easier when it comes to consistency. Collective beliefs are only attributed to subgroups, which may be so chosen that individual beliefs are not too dissimilar. But the task is in another respect harder because the constitution of (sub)groups is endogenized. It is now part of the process of belief attribution.

In what follows, I shall be concerned only with clustering and aggregation, leaving aside the preliminary step consisting in the identification of the dimensions which structure the issue under scrutiny. The omission is not harmful. It is an optional step, which is partly motivated by pragmatic considerations. Data-reduction techniques make it easier to get an intuitive grasp of the clusters. Moreover, one may simply assume that clustering and aggregation, as they are going to be described, operate on data that has already been organized around structuring dimensions. But getting a better understanding of the role played by structuring dimensions in doxastic clustering is certainly of interest. We leave it to further work.

The setting is mostly borrowed from judgment aggregation theory. Issues under scrutiny are represented by the agenda. An agenda A is a set of formulas closed under negation. Individual opinions are represented by judgment sets. A judgment set J is a coherent and complete subset of A. Opinions in the sample are represented by a profile. A profile P is a sequence of judgment sets of finite length. $\mathcal{J}$ is the set of all judgment sets and $\mathcal{P}$ is the set of all profiles. Agents are identified with integers so the judgment set of agent i is $P[i]$, the ith element of P. Note that the number of agents is not fixed in advance.

Judgment aggregation deals with aggregation functions that map profiles onto subsets of the agenda. Our focus of interest will be *clustering methods*, which perform in one swoop the last two steps we aim at modeling. They define subgroups within the sample and at the same time they attribute beliefs to these subgroups. Formally, a clustering method $C: \mathcal{P} \rightarrow \mathcal{P}$ is simply a function transforming a profile into another profile of the same length. Thus, a clustering method transforms the opinions $P[i]$ of an individual i into opinions $C(P)[i]$. But $C(P)[i]$ is to be interpreted as something quite

[4] This final remark is merely meant to be suggestive. A detailed account of the ontological matters regarding the existence and nature of doxastic groups as social objects of a definite kind goes beyond the scope of the present chapter.

different from $P[i]$. It is to be thought as the opinions of the cluster i belongs to. Now clusters do not need to be an explicit output of the clustering method, because they have just been simply implictly defined. Given a profile P and a clustering method C, agents i and j belong to the same cluster iff $C(P)[i] = C(P)[j]$. Belonging to the same cluster is an equivalence relation $\sim_{C,P}$ whose equivalence classes are the C-clusters in P. Thus I is a cluster obtained through the clustering method C in the profile P iff $I \in P_{/\sim_{C,P}}$.[5] We may talk about the judgment set $C(P)[I]$ of cluster I in P, since $C(P)[i] = C(P)[i']$ for all $i, i' \in I$. Given C, we also say that a judgment set J determines a cluster I, where I is $\{i \in P \mid C(P)[i] = J\}$, for $I \neq \emptyset$. In what follows, I assume that C is a total function, which corresponds to the condition of universal domain in judgment aggregation. This condition ceases to be problematic in so far as the group can be split in as many clusters as needed.

8.6 Axioms for Doxastic Clustering

We shall now consider some desirable properties of clustering methods. First of all, clustering methods should be robust when the group is extended in certain specific ways. In general, clustering is not a conservative endeavor. Enlarging the sample might result in clusters on the bigger sample whose restriction to the smaller sample does not match the previous clusters on the smaller sample. But the clustering should remain unchanged when new agents are added to the group, whose doxastic profile *confirms* the clustering that had previously been performed. It is particularly natural to consider such extensions when one deals with groups that are samples taken within a larger population.

Let us note $P \cdot J$ the extension of sequence P by the judgment set J, which corresponds to adding an agent to the group with opinions J. Our first robustness requirement says that adding an agent whose judgments exactly correspond to the judgments of one cluster should leave the clustering unchanged. Such an agent is incorporated to its natural cluster. Intuitively, if we encounter a new agent whose opinions exactly correspond to the kind of opinions we take as typical of some group in our sample, this should not change our views regarding which opinions are typical. Clustering should be stable with respect to such an extension:

Definition 1 (Stability). *A clustering method C is stable iff, for all $P \in \mathcal{P}$, $J \in \mathcal{J}$,*

$$\textit{if there is } i \in n \textit{ and } J \in \mathcal{J} \textit{ such that } J = C(P)[i] \textit{ then } C(P \cdot J) = C(P) \cdot J$$

Secondly, we might take another sample of the same size in our population. If we get lucky, opinions in this new sample are the same as opinions in our first sample. In

[5] Our framework and our notations are a, hopefully not too confusing, mix between what is familiar in judgment aggregation theory and in the clustering literature. In the clustering literature, clustering methods typically output a set $C = \{C_1, ..., C_k\}$ of clusters. C is a partition of an unstructured data set equipped with a distance function. Our data set is structured, consisting of a set of judgment sets. In terms of notations, C denotes in the present chapter the clustering method and not just its ouput. And the output is not just the clusters but rather the clusters *and* the information best summarizing what each cluster thinks.

this case, we again have no reason to revise our clustering. Clustering should then be invariant under uniform duplication of agents. This requires our methods to be scalable:

Definition 2 (SCALABALITY). *A clustering method C is scalable iff, for all $P \in \mathcal{P}$,*

$$C(P \cdot P) = C(P) \cdot C(P)$$

In some early work on general properties of clustering methods, Fisher & van Ness (1971) considered requiring that the 'geometrical shape of clusters' (1971, p. 96) is preserved when the density of the domain is altered by adding more points, identical to some of the points which were already there. Stability and scalability are then special cases of this property of 'point proportion admissibility,' with the proviso that the points added in the stability condition may not have been actually present in the domain.

Now we are considering *doxastic* clusterings. Clustering and aggregation should only depend on the doxastic states of agents and not on who these agents are. Discarding any extra-information associated with the identity of agents is captured through the familiar requirement of anonymity:

Definition 3 (ANONYMITY). *A clustering method C is anonymous iff, for all $P \in \mathcal{P}$, for all permutations π on P,*

$$C(\pi(P)) = \pi(C(P))$$

The label 'anonymity' is familiar from judgment aggregation theory, whereas the literature on clustering usually speaks in terms of isomorphism invariance (Ackerman, Ben-David & Loker, 2010a).

Finally, we lay out some requirements on intra-cluster aggregation that are straightforwardly adapted from judgement aggregation theory. Since we are dealing with unorganized groups and accept the strong supervenience thesis, we need to limit the information used in deciding on a given belief for a given cluster to the corresponding beliefs of the members of that cluster, in a way that does not depend on which belief is considered. Let us first introduce the technical notion of support for a formula in a group of agents. Given a profile P, the support of a formula ϕ in a subgroup of agents I, which we note $S_P(I,\phi)$, is the set of agents in I who think that ϕ, that is $S_P(I,\phi) = \{i \in I \mid \phi \in P[i]\}$. We can now state the following adaptation of the familiar condition of systematicity:

Definition 4 (SYSTEMATICITY). *A clustering method C is systematic iff, for all $\phi, \psi \in X$, $P, P' \in \mathcal{P}$, and for all subgroups $I \subseteq N$ such that I is a C-cluster in P and also a C-cluster in P',*

$$\text{if } S_P(I,\phi) = S_{P'}(I,\psi) \text{ then } \phi \in C(P)[I] \text{ iff } \psi \in C(P')[I]$$

Furthermore, we require that unanimity within a cluster be respected.

Definition 5 (UNANIMITY). *A clustering method C is unanimous iff, for all $P \in \mathcal{P}$, for all C-clusters I in P,*

$$\text{if } S_P(I,\phi) = I \text{ then } \phi \in C(P)[I]$$

If we were dealing with aggregation functions, the last three requirements could not be jointly satisfied.[6] The situation with clustering methods is very different; inconsistencies can be avoided simply by avoiding to make up clusters in which a Condorcet paradox would occur. Consider the identity function, which is the trivial clustering method $Id : \mathcal{P} \to \mathcal{P}$ defined by $Id(P) = P$. *Id* puts two agents in the same cluster if and only if they have the same beliefs, and the cluster collective opinion is simply the unanimous opinion of its members. It is easy to see that *Id* satisfies systematicity, unanimity, and anonymity, as well as stability and scalability. In order to know more about clustering methods that satisfy these requirements, we shall prove in the next section a theorem characterizing their consequences on intra-cluster aggregation.

8.7 A Theorem about Distance Minimization

Clustering on a set of objects typically uses a distance relation on these objects. Loosely speaking, objects which are close should belong to the same cluster, and objects which are not should not. With respect to intra-cluster aggregation, a desirable property of collective opinions for a cluster is that they minimize distance with respect to individual opinions of its members. Formally, a distance on $\mathcal{P}$ is a function $d : \mathcal{P} \times \mathcal{P} \to \mathbb{R}^+$ such that for all $P, P' \in \mathcal{P}$, $d(P,P') = d(P',P)$, $d(P,P') = 0$ iff $P = P'$ and for all $P, P', P'' \in \mathcal{P}$, $d(P,P'') \leq d(P,P') + d(P',P'')$. Within clusters, distance minimization is then defined in the following way:

Definition 6 (DISTANCE MINIMIZATION). *Let C be a clustering method and d a distance on profiles, C minimizes distances with respect to d iff, for all $P \in \mathcal{P}$, for all C-clusters I in P, for all $K \subseteq X$,*

$$\sum_{i \in I} d(P[i], K) \leq \sum_{i \in I} d(P[i], C(P)[I]) \textit{ implies } K = C(P)[I]$$

The simplest distance counts the number of differences between two profiles. It is the so-called Hamming distance $h : \mathcal{P} \times \mathcal{P} \to \mathbb{R}^+$ which is defined by

$$h(P,P') = |\{\phi \in X \mid \phi \in P \text{ and } \phi \notin P'\} \cup \{\phi \in X \mid \phi \notin P \text{ and } \phi \in P'\}|$$

We are now ready to state the main theorem, which shows that minimization of Hamming distance is exactly what systematicity and unanimity require, given stability, scalability, and anonymity as background assumptions on clustering methods.

Theorem 7. *Let C be a stable, scalable and anonymous clustering method, C minimizes Hamming distance iff C is systematic and unanimous.*

The proof of Theorem 7 partly relies on two lemmas which are already known in a slightly different form in judgment aggregation theory.

[6] To be precise, two more conditions are needed for the classical impossibility theorem, namely universal domain and collective rationality. But note that these two conditions are built in in our framework. Clustering methods are required to be total functions, which corresponds to universal domain, and they output profiles, which enforces collective rationality at the level of clusters.

Lemma 8. *A clustering method C minimizes Hamming distance iff for all profiles P, for all C-clusters I in P, for all $\phi \in X$,*

$$\phi \in C(P)[I] \text{ iff } |S_P(I,\phi)| > |S_P(I,\neg\phi)|$$

Lemma 9. *If a clustering method C satisfies Systematicity and Anonymity, then for all profiles P, for all C-clusters I in P, for all $\phi, \psi \in X$,*

$$\text{if } |S_P(I,\phi)| = |S_P(I,\psi)|, \text{ then } \phi \in C(P)[I] \text{ iff } \psi \in C(P)[I]$$

Lemma 8 says that minimization of Hamming distance within clusters is simply majoritarian aggregation within clusters. Note that the Hamming distance is by no means the only sensible distance which might be used to compare judgment sets, even though it is probably the simplest choice in the present context. But then Lemma 8 may be read as a further argument in favor of this choice. It is reasonable to assume that what a cluster thinks should correspond both to its center, in the sense of the point which minimizes the sum of distances to all points in the cluster, and to the views which are upheld by a majority of people in the cluster. Lemma 8 says that the two characterizations coincide when the underlying distance is the Hamming distance. Lemma 9 is more technical in nature and says that jointly requiring systematicity and anonymity makes aggregation within clusters a matter of size.

The Left to Right direction of Theorem 7 directly follows from Lemma 8 and the fact that majoritarian aggregation is systematic and anonymous. The direction from Right to Left is congenial to the clustering framework, using both Lemmas 8, 9, and the clustering specific axioms of scalability and stability. Detailed proofs of Theorem 7 and Lemmas 8 and 9 are given in the Appendix.

Theorem 7 uniquely characterizes an aggregation procedure within clusters. It does not uniquely characterize a clustering method. Two things are packed into our definition of clustering methods: the clustering strictly speaking, which consists in partitioning the set of agents, and intra-cluster aggregation, which consists in attributing to each cluster a judgment set. As defined, minimization of Hamming distance within clusters concerns intra-cluster aggregation.[7] Presumably, it is compatible with lots of different ways of clustering agents. As an example, we have seen that the trivial clustering method *Id* satisfies all our axioms. So would clustering methods based on *k*-means, provided clusters for which majoritarian voting results in a Condorcet paradoxes are excluded. Another method satisfying the axioms would be to first pick up the largest subgroup which is unidimensional, and then the second largest subgroup which is so, and so on, as suggested in a different context by List (2001). To fully take advantage of the clustering framework, further research should combine characterizations of clustering and characterizations of intra-cluster aggregation. When intra-cluster aggregation relies on a distance, a natural bridge principle would

[7] Partitioning and intra-cluster aggregation are not independent. Because of the requirement that intra-cluster aggregation yields coherent and complete subsets of the agenda, the choice of an intra-cluster aggregation function constrains the set of available clusters.

be to require that the distances used for clustering and for intra-cluster aggregation are the same.

Theorem 7 may be compared with characterizations of majoritarian voting in judgment aggregation theory. In favorable circumstances, that is for unidimensional profiles,[8] systematicity and anonymity uniquely characterize majoritarian voting for sufficiently rich agendas (List, 2003). In the clustering framework, there is no need for well-behaved profiles, since the clustering method is to perform the necessary splitting into subgroups. In List's Theorem and in Theorem 7, systematicity and anonymity play a similar role. They make aggregation a matter of cardinalities. The main difference lies in the extra step needed to show that matters of cardinalities actually boil down to majority voting. Given that, in both frameworks, the output is to be coherent and complete, proving monotonicity is enough, because then the cutoff will have to be at half-half.[9] The proof of monotonicity in the judgment aggregation case depends on the behavior of the aggregation function on logically connected formulas in the agenda, which is why assumptions on the agenda are made. Such assumptions are not needed, and would not work, in the clustering framework. They would not work because logical pressure can be ignored by means of further splitting. They are not needed because properties congenial to clustering, namely stability and scalability, force intra-cluster aggregation to be well behaved and, in particular, monotonic.

Finally, leaving technicalities aside, how exactly is Theorem 7 relevant to the question of the existence and of the nature of collective opinions in unorganized groups? According to the view defended in Section 8.2, collective opinions in unorganized groups should satisfy strong supervenience. Following Section 8.3, strong supervenience constrains the possible patterns of dependence of collective opinions on individual opinions, in ways that are reflected by the formal conditions of anonymity and independence. On this basis, the classical impossibility result of judgment aggregation theory threatens the very existence of collective opinions in unorganized groups. In Section 8.4, it was argued that collective opinions in unorganized groups make sense (only) for doxastically homogenous subgroups. This suggests a response to the threat caused by the classical impossibility result: maybe what is impossible when the group is taken as a given becomes possible when the constitution of groups to which collective beliefs are attributed depend on what their members think. Theorem 7 proves the intuition to be correct. Even under mild assumptions on the robustness of the definition of subgroups, there are clustering methods satisfying the constraints associated with strong supervenience.[10] Moreover, these constraints yield majoritarian aggregation as the natural notion of collective beliefs for such groups. Thus, collective opinions in unorganized groups are just what we think they are: average opinions that correspond to some coherent or homogenous way of thinking about an issue.

[8] Unidimensionality is the best known condition, but majoritarian voting is consistent for some wider classes of profiles. See Dietrich and List (2010) for results showing how to go beyond unidimensionality.

[9] See the first part of the proof of Theorem 7 in the Appendix.

[10] Note that for technical reasons, Theorem 7 relies on systematicity rather than independence.

8.8 Conclusion

Formal, as well as conceptual, studies on collective opinion usually presuppose that a group is given, and the problem is then taken to be to understand belief attribution to this constituted group, or how beliefs of individuals in this constituted group may be aggregated. This way of framing the problem is not always adequate. Whenever we are dealing with loose communities rather than well-structured groups, collective opinions come in many colors. What matters is to identify kinds of opinions, and how these kinds of opinions are distributed in the community, forming clusters of likely minded individuals. This paves a new way for the formal study of judgment aggregation. Rather than taking the group as an unbreakable doxastic unit, models should account for what the relevant doxastic units are. They should deal both with clustering and with aggregation of individual beliefs.

The final sections of the present chapter have proposed a framework and some axioms that may be used to provide a formal analysis of this joint process. But, obviously, much remains to be done. In particular, we are saying that clusters are to consist of like-minded individuals. But how is doxastic homogeneity in subgroups to be evaluated? How are the subgroups built? In order to answer this question, one would need to look into further desirable properties of clustering methods, bearing upon what should guide the process of grouping together some individual opinions rather than some others. In this respect, the robustness axioms in Section 8.6 and the theorem on distance minimization in Section 8.7 are merely preliminary results suggesting directions for further work.

Since doxastic clustering is at the crossroads of clustering and aggregation, the general plan is to unify research on clustering and research on judgment aggregation. Let us say a few words about the former and what the prospects for the latter are. Following Kleinberg (2002) impossibility results, recent research in statistics has focused on providing axiomatic classifications of the variety of existing clustering methods (Ackerman & Ben-David, 2008, Ackerman, Ben-David & Loker, 2010a and 2010b) and unique characterizations of some particular clustering methods have been given (e.g., Zadeh & Ben-David, 2009 for single linkage). By contrast, Theorem 7 did not aim at selecting some specific clustering methods for the clustering of judgment sets. It only investigates the effects of the standard axioms from the theory of judgment aggregation, in a context where some algorithm respecting the rather weak stability and scalability requirements is assumed to generate clusters of opinions.[11] Further research should investigate the effects of adding stronger clustering axioms to constrain how opinions are grouped together, such as locality requirements (when the input is restricted to the points belonging to some of the generated clusters, the clustering method still outputs the same clusters), richness conditions (the clustering

[11] This difference in focus results in technical differences regarding the formal frameworks. In the literature on clustering, a distance or similarity function between objects is given, whereas Theorem 7 characterizes the distance to be used in summarizing the information in each cluster. But in some other respects, our approach is the more narrow one. In particular, we have set aside the question how an optimal number of clusters is determined and the possibility of producing a hierarchical tree rather than a single clustering solution.

method should be able to generate any partition when distances between points are suitably modified) or stronger forms of scale invariance (e.g., clusters do not change when distances undergo a monotone transformation). By so doing, one may hope to get the best out of the domain-specific stance outlined here and the general approach developed in recent works on clustering axioms.

Acknowledgments

I would first like to thank Franz Dietrich for pointing out a mistake in an earlier version of Theorem 7 and its proof (indeed, using independence rather than systematicity does not work). I am indebted to J. Baccelli, M. Cozic, H. Galinon, and B. Simmenauer for enjoyable discussions and stimulating remarks around this project on doxastic clustering. I also benefited from two particularly enlightening, and encouraging, referee reports. Finally, I gratefully acknowledge support from the Labex and Idex grants: ANR-10-LABX-0087 IEC and ANR-10-IDEX-0001-02 PSL.

References

Ackerman, M. & Ben-David, S. (2008) "Measures of Clustering Quality: A Working Set of Axioms for Clustering," *Advances in Neural Information Processing Systems*, 21, pp. 121–128.

Ackerman, M., Ben-David, S. & Loker, D. (2010a) "Towards Property-based Classification of Clustering Paradigms," *Neural Information Processing Systems Conference* (NIPS 2010).

Ackerman, M., Ben-David, S. and Loker, D. (2010b) "Characterization of Linkage-based Clustering," *Conference on Learning Theory* (COLT 2010), pp. 270–281.

Dietrich, F. & List, Ch. (2007) "Arrow's Theorem in Judgment Aggregation," *Social Choice and Welfare*, 29, pp. 19–33.

Dietrich, F. & List, Ch. (2010) "Majority Voting on Restricted Domains," *Journal of Economic Theory*, 145, pp. 512–543.

Dokow, E. & Holzman, R. (2010) "Aggregation of Binary Evaluations," *Journal of Economic Theory*, Symposium on Judgment Aggregation, 145, pp. 495–511.

Fisher, L. & van Ness, J.W. (1971) "Admissible Clustering Procedures," *Biometrika*, 58 (1), pp. 91–104.

Gilbert, M. (1989) *On Social Facts*, Routledge.

Kleinberg, J. (2002) "An Impossibility Theorem for Clustering," *Advances in Neural Information Processing Systems*, 15, pp. 446–453.

List, Ch. (2001) *Mission Impossible? The Problem of Democratic Aggregation in the Face of Arrow's Theorem*, DPhil-thesis in Politics, University of Oxford.

List, Ch. (2003) "A Possibility Theorem on Aggregation over Multiple Interconnected Propositions," *Mathematical Social Sciences*, 45, pp. 1–13.

List, Ch. (2004) "A Model of Path-Dependence in Decisions over Multiple Propositions," *American Political Science Review*, 98, pp. 495–513.

List, Ch. & Pettit, Ph. (2002) "Aggregating Sets of Judgments: An Impossibility Result," *Economics and Philosophy*, 18, pp. 89–110.

List, Ch. & Pettit, Ph. (2011) *Group Agency*, Oxford University Press.

Pettit, Ph. (2003) "Groups with Minds of Their Own," *in* F. Schmitt (ed.), *Socializing Metaphysics: The Nature of Social Reality*, Rowman & Littlefield, pp. 167–93.

Quinton, A. (1975) "Social Objects," *Proceedings of the Aristotelian Society*, 76, pp. 1–27.

Zadeh, R.B. & Ben-David, S. (2009) "A Uniqueness Theorem for Clustering," *in Proceedings of the Twenty-Fifth Conference on Uncertainty in Artificial Intelligence* (UAI2009), pp. 639–646.

Appendix

First, we prove the two lemmas used in the proof of Theorem 7.

Lemma 8. *A clustering method C minimizes Hamming distance iff for all profiles P, for all C-clusters I in P, for all $\phi \in X$,*

$$\phi \in C(P)[I] \textit{ iff } |S_P(I,\phi)| > |S_P(I,\neg\phi)|$$

Proof. From Left to Right. Let C be a clustering method which minimizes Hamming distance. Assume for contradiction that there is a profile P, a C-cluster I in P and a formula ϕ such that $\phi \in C(P)[I]$ and $|S_P(I,\phi)| \leq |S_P(I,\neg\phi)|$. Let $K = (C(P)[I] \setminus \{\phi\}) \cup \{\neg\phi\}$. By definition of h,

$$\sum_{i \in I} h(P[i],K) = \sum_{i \in I} h(P[i],C(P)[I]) - 2 \cdot |S_P(I,\neg\phi)| + 2 \cdot |S_P(I,\phi)|$$

Hence, since $|S_P(I,\phi)| \leq |S_P(I,\neg\phi)|$,

$$\sum_{i \in I} h(P[i],K) \leq \sum_{i \in I} h(P[i],C(P)[I])$$

Since $K \neq C(P)[I]$, this contradicts the fact that C minimizes Hamming distance.

From Right to Left. Assume that C is a clustering method such that for all profiles P, for all C-clusters I in P, $\phi \in C(P)[I]$ iff $|S_P(I,\phi)| > |S_P(I,\neg\phi)|$. Given a profile P and a C-cluster I in P, let $D = \sum_{i \in I} h(P[i],C(P)[I])$. It is sufficient to show that, for any $K \subseteq X$ with $K \neq C(P)[I]$,

$$\sum_{i \in I} h(P[i],K) > D$$

Now let $D^- = C(P)[I] \setminus K$ and $D^+ = K \setminus C(P)[I]$. By definition of h,

$$\sum_{i \in I} d(P[i],K) = D + \sum_{\phi \in D^-} (|S_P(I,\phi)| - |S_P(I,\neg\phi|)) + \sum_{\phi \in D^+} (|S_P(I,\neg\phi)| - |S_P(I,\phi)|)$$

By the assumption on C, for any $\phi \in D^-$, $|S_P(I,\phi)| > |S_P(I,\neg\phi|$, and for any $\phi \in D^+$, $|S_P(I,\neg\phi)| > |S_P(I,\phi|$, therefore $\sum_{i \in I} h(P[i],K) \geq D$. Moreover, since $K \neq C(P)[I]$, at least one of D^- and D^+ is non-empty, so the inequality is strict. □

Lemma 9. *If a clustering method C satisfies Systematicity and Anonymity, then for all profiles P, for all C-clusters I in P, for all $\phi, \psi \in X$,*

$$\textit{if } |S_P(I,\phi)| = |S_P(I,\psi)|, \textit{ then } \phi \in C(P)[I] \textit{ iff } \psi \in C(P)[I]$$

Proof. Let C be a clustering method satisfying systematicity and anonymity. Let P be a profile, I a C-cluster in P, ϕ, ψ two formulas in the agenda such that $|S_P(I,\phi)| = |S_P(I,\psi)|$. Let π be a permutation on P such that $\pi(S_P(I,\phi)) = S_P(I,\psi)$ and π is the identity outside $S_P(I,\phi) \cup S_P(I,\psi)$. By definition of π, $S_P(I,\phi) = S_{\pi(P)}(I,\psi)$, hence, by systematicity, $\phi \in C(P)[I]$ iff $\psi \in C(\pi(P))[I]$. By anonymity, I is a C-cluster in $\pi(P)$ and $C(P)[I] = C(\pi(P))[I]$. Therefore $\phi \in C(P)[I]$ iff $\psi \in C(P)[I]$ as required. □

Finally, we prove the Right to Left direction of Theorem 7.

Theorem 7. *Let C be a stable, scalable and anonymous clustering method, C minimizes Hamming distance iff C is systematic and unanimous.*

Proof. Let C be a stable, scalable, and anonymous clustering method. Assume that C is systematic and unanimous, and show that C minimizes Hamming distance. Consider an arbitrary profile P and any C-cluster I in P. By Lemma 8, it is sufficient to show that, for all $\phi \in X$, $\phi \in C(P)[I]$ iff $|S_P(I,\phi)| > |S_P(I,\neg\phi)|$. Let $D_I = \{m \in \mathbb{N} \mid \text{there is } \phi \in X \text{ with } |S_P(I,\phi)| = m\}$. Since C is systematic and anonymous, by Lemma 9, there is a unique set $W_I \subseteq D_I$ such that, for all $\phi \in X$, $\phi \in C(P)[I]$ iff $|S_P(I,\phi)| \in W_I$. Hence, it is sufficient to show that $W_I = \{m \in D_I \mid m > |I|/2\}$.

Say that W_I is closed upward if $m \in W_I$ and $m' > m$ with $m' \in D_I$ implies $m' \in W_I$. Showing that W_I is closed upward completes the proof. This is because, in that case, there is a $k \in \mathbb{N}$ such that $m \in W_I$ iff $m \in D_I$ and $m > k$. By coherence and completeness of $C(P)[I]$, this k can be taken to be equal to $\lfloor |I|/2 \rfloor$. To see this, first note that, when $|I|$ is even, $|I|/2 \notin D_I$, because consistency would force $|I|/2$ out of W_I and completeness would force $|I|/2$ into W_I. Since $C(P)[I]$ is consistent and W_I is closed upwards, $m \in W_I$ implies $m > \lfloor |I|/2 \rfloor$. Moreover, since $C(P)[I]$ is complete, $m \in D_I$ and $m \geq \lfloor |I|/2 \rfloor$ implies $m \in W_I$.

Finally, we show that W_I is closed upward. Let P be such that there is C-cluster I in P and formulas $\phi, \psi \in X$ with $|S_P(I,\phi)| = m$, $|S_P(I,\psi)| = m'$ and $\phi \in C(P)[I]$. We need to show that $\psi \in C(P)[I]$ as well. Let $|I| = n$. If $m' = n$, we immediately get $\psi \in C(P)[I]$ by unanimity. Assume $m' < n$. By Scalability,

$$C(\underbrace{P \cdot P \ldots \cdot P}_{\text{n-m' times}}) = \underbrace{C(P) \cdot C(P) \ldots \cdot C(P)}_{\text{n-m' times}}$$

Let K be short for $C(P)[I]$. K determines a cluster I_2 of size $n \cdot (n - m')$ in the larger sample $P \cdot P \ldots \cdot P$. Moreover, $|S_{P \cdot P \ldots \cdot P}(I_2,\phi)| = m \cdot (n - m')$, since each $C(P)$ contributes m agents thinking that ϕ to cluster I_2. By stability,

$$C(\underbrace{P \cdot P \ldots \cdot P}_{\text{n-m' times}} \cdot \underbrace{K \cdot \ldots \cdot K}_{\text{n} \cdot \text{(m'-m) times}}) = C(\underbrace{P \cdot P \ldots \cdot P}_{\text{n-m' times}}) \cdot \underbrace{K \cdot \ldots \cdot K}_{\text{n} \cdot \text{(m'-m) times}}$$

K now determines a cluster I_3 of size $n \cdot (n - m') + n \cdot (m' - m)$ in the larger sample $P \cdot P \ldots \cdot P \cdot K \cdot \ldots \cdot K$. Moreover, $|S_{P \cdot P \ldots \cdot P \cdot K \cdot \ldots \cdot K}(I_3,\phi)| = m \cdot (n - m') + n \cdot (m' - m)$, since each K contributes one agent thinking that ϕ to cluster I_3.

Similarly, apply scalability and get

$$C(\underbrace{P \cdot P \ldots\ldots \cdot P}_{\text{n-m times}}) = \underbrace{C(P) \cdot C(P) \ldots\ldots \cdot C(P)}_{\text{n-m times}}$$

K determines a cluster I_4 of size $n \cdot (n - m)$ in the larger sample $P \cdot P \ldots\ldots \cdot P$. Moreover, $|S_{P \cdot P \ldots\ldots \cdot P}(I_4,\psi)| = m' \cdot (n - m)$, since each $C(P)$ contributes m' agents thinking that ψ to cluster I_4.

Now, $n \cdot (n - m') + n \cdot (m' - m) = n \cdot (n - m)$, hence I_3 and I_4 have the same size, and there exists a permutation π on $P \cdot P \ldots\ldots \cdot P$ such that $I_3 = \pi(I_4)$. By anonymity,

$$C(\pi(\underbrace{P \cdot P \ldots\ldots \cdot P}_{\text{n-m times}})) = \pi(\underbrace{C(P) \cdot C(P) \ldots\ldots \cdot C(P)}_{\text{n-m times}})$$

Hence K determines the cluster I_3 in the sample $\pi(P \cdot P \ldots\ldots \cdot P)$. As before, $|S_{\pi(P \cdot P \ldots\ldots \cdot P)}(I_3, \psi)| = m' \cdot (n - m)$. But $m' \cdot (n - m) = m \cdot (n - m') + n \cdot (m' - m)$, hence

$$|S_{\pi(P \cdot P \ldots\ldots \cdot P)}(I_3, \psi)| = |S_{P \cdot P \ldots \cdot P \cdot K \cdot \ldots \cdot K}(I_3, \phi)| \quad (*)$$

The C-clusters I_3 in $\pi(P \cdot P \ldots\ldots \cdot P)$ and I_3 in $P \cdot P \ldots \cdot P \cdot K \cdot \ldots \cdot K$ have the same collective beliefs K. By hypothesis, $\phi \in K$, hence, by Lemma 9 and the Equality $(*)$, $\psi \in K$, as required. □

9

Opinion Aggregation and Individual Expertise

CARLO MARTINI AND JAN SPRENGER

9.1 Introduction

Members of a group are often in **disagreement** with each other. Analysts at Apple or Samsung come up with different estimates of how often a newly developed cell phone will be sold. Conservation biologists disagree on the population size of an endangered species. Researchers at the European Central Bank cannot find a consensus on the merits and drawbacks of a particular monetary policy. In all these cases, disagreement need not be a bad thing. What is more, it can be seen as epistemically and socially desirable. Armstrong (2001) and Page (2007), among others, argue that the diversity of opinion characteristic of disagreement can act as an antidote to groupthink and foster the development of alternative approaches whose pursuit may be more fruitful than following well-trodden paths.

On the other hand, disagreement can block the formation of group judgment and delay important decisions. To stay with one of the above examples, the European Central Bank needs to come up with *some* decision on whether to ease or to tighten monetary policy. Also in science, it seems that a certain degree of consensus is necessary for scientific progress and for conducting "normal science" (Kuhn 1962, 1977). So how should groups **aggregate (or pool) the opinions of their members**?

This question has no clear-cut, unique answer. One reason is the diversity of contexts where groups aggregate individual opinions, and the different goals they aim at. Another reason is the diversity of criteria for evaluating aggregation procedures, e.g., epistemic and social ones. In the epistemic perspective, opinion aggregation procedures are primarily assessed according to their veracity, that is, their success at tracking the truth. In the social perspective, they are judged on different criteria: Was the opinion of every individual duly taken into account? Is the final result acceptable to all group members? Did it emerge from a procedure that everybody consented to? And so on. To give a crude example, a dictatorship is usually ruled out in the social perspective, but it may be acceptable from an epistemic perspective, if that individual is by far the most competent group member.

Boyer-Kassem, Mayo-Wilson, & Weisberg. *Scientific Collaboration and Collective Knowledge*.

There are two principled ways of aggregating opinions and resolving disagreements: with and without **belief revision on an individual level**. The main part of the chapter is devoted to models of opinion aggregation that do not involve belief revision. Such models describe in a formal framework how opinion aggregation procedures should (or factually do) work.[1] However, we will also contrast them with consensual decision-making where individuals revise their beliefs and eventually agree on the subject matter of disagreement. Typical driving factors of such procedures are deliberation and considerations of power and mutual respect.

It is useful to divide opinion aggregation models that do not involve deliberation or belief revision into two categories. In the first category are **egalitarian models**—a term supposed to capture the fact that no individual has a special or privileged position. They are particularly important in contexts where it is hard to argue for giving special weight to a specific individual, e.g., because relevant expertise is hard to elicit. Sometimes, it is even one of the constraints on an aggregation procedure that the opinions be "anonymous," that is, that the final outcome (the group opinion) do not depend on the individual agents that submitted a view. This aggregation procedure is acting like the famous allegory of justice: It weighs opinions impartially and regardless of the agent's identities.

The use of egalitarian models may have epistemic, procedural, or pragmatic reasons. For example, the equality of individual votes in elections or referenda—a particular case of social opinion aggregation—has constitutional status in most democracies. Therefore, choosing an egalitarian procedure may be pragmatically required. Hence, it comes as no surprise that egalitarian aggregation procedures, such as various forms of majority or plurality voting, play a major role in political processes.

The second category of aggregation models—which we will call **differential models**—gives differential weights to the individual group members. This procedure can have psychological as well as epistemic motivations: Firstly, "fringe opinions"—i.e. those that are far away from the group average—may receive low weight precisely because they appear to be wide off the mark, and nobody else is willing to take them seriously. Secondly, when the task is *intellective*, that is, when it involves a high level of demonstrability, some group members may be more competent than others. In such tasks, the opinions of the **experts** might receive higher consideration in determining the group view. Here we think of experts as the more competent group members, that is, those whose opinion coincides more often with the truth, or contains the smallest error (see Goldman 1999). Thirdly, there are group members that may, because of their position or their appearance, receive more respect or esteem than others in the peer-group, and it may be procedurally or pragmatically necessary to endow them with a higher weight than their peers.

Which of all these procedures is applied in practice is a context-sensitive issue: For instance, in a meeting of the heads of state of the European Union on foreign policy or the EU budget, the primary goal may consist in finding a consensual position, and that need not necessarily involve egalitarian decision-making. Also when the epistemic

[1] We use "procedure" as a generic word for ways to aggregate opinions and to resolve disagreements, whereas "model" denotes a particular formal framework for describing these procedures.

accuracy of the group judgment has the highest priority, e.g., in scientific reasoning, it may be rational to weigh differentially and to defer to experts. More generally, List (2005) distinguishes between two challenges for group judgments: the "rationality challenge" of endorsing a consistent collective judgment on an interconnected agenda of propositions, and the "knowledge challenge," that is, to track truth in these collective judgments. Both challenges will be discussed in this chapter.

Finally, we should distinguish between weighting *strategies* and weighting *schemes*. A strategy is general: In the classification presented in this chapter, it is either egalitarian or differential. A scheme is specific: It is clear that once a differential weighting strategy has been selected, a specific set of weights (a scheme) is needed. Perhaps contrary to our intuitions, egalitarian aggregation procedures can allow for different weighting schemes. Equal-weight views in the epistemology of disagreement literature are often vague on the issue of which specific function aggregates the individual opinions; for example, the "equal-weight view" is sometimes interpreted in a Bayesian scheme (Elga 2007) and other times as linear averaging (Christensen 2009).

In sum, this chapter sketches a rough map of diverse approaches to aggregating group opinions that matter for science, politics, and social decision-making more generally. Particular attention will be devoted to differential models, and among them, to models where an agent's weight is determined by his or her expertise in the subject area. These models are contrasted with egalitarian or power-based opinion aggregation. Notably, we omit the problem of information aggregation and assume that all agents are on a par with respect to the available information. Possible differences between them are thus a matter of different judgmental prowess. For reasons of space, we also have to ignore the problems related to the value of disagreement itself (in science and society) against the demands for consensus; the interested reader will find useful discussions in Beatty and Moore (2010) and Beatty (2006).

The two following sections are devoted to the problem of group rationality in opinion aggregation: how should a group aggregate individual opinions if constrained by a set of epistemic or social requirements? While Section 9.2 deals with aggregating **binary propositional attitudes**, that is, yes-no judgments, Section 9.3 investigates the aggregation of **graded attitudes**, and in particular, probability distributions. Sections 9.4 and 9.5 both describe the problem of aggregating **numerical estimates**, but from different perspectives. First, we examine philosophical and empirical motivations for **consensual belief revision** in the light of disagreement and review a major formal model of such a procedure, the Lehrer-Wagner model. Then, we conduct an analysis of the epistemic benefits of differential weighting as opposed to straight averaging. Section 9.6 gives an overview of empirical findings on expertise, e.g., the identification and ranking of experts, and relates them to the formal results. Section 9.7 concludes.

9.2 Judgment Aggregation

Judgment aggregation is a recent field at the intersection of philosophy and economics. It deals with the aggregation of individual judgments on an interconnected set of propositions into a coherent group judgment and asks what kind of aggregation procedure rationality requires us to choose. For our purposes, judgment aggregation is particularly interesting because procedural constraints may require democratic decision-making, where social and epistemic features of individual agents are not taken into account. Judgment aggregation procedures are typically egalitarian: no single person obtains more weight than others because of his/her standing in the group or his/her competence. Studying judgment aggregation, and its problems and methods, also yields a better understanding of the advantages and drawbacks of differential weighting in opinion aggregation.

Notably, it has been shown that the classical economic problem of preference aggregation can be embedded into the problem of judgment aggregation (Dietrich and List 2008; Grossi 2009). Classical results such as Arrow's impossibility theorem can be represented as impossibility theorems for aggregating judgments on logically interconnected propositions. This increases the generality and relevance of theoretical results on judgment aggregation.

Let us now formulate the classical problem of judgment aggregation. Assume that N agents are supposed to aggregate their binary judgments on an agenda of propositions $X = \{A_1, \ldots, A_m\}$ where the A_k can be logically interconnected. Call J_i the judgment set of agent i. Then we can ask ourselves which kind of condition should be satisfied by an aggregation function $F : \mathcal{J}^n \rightarrow \mathcal{J}$ that maps the individual agents' judgment sets to a group judgment set $F(J_1, \ldots, J_n)$.

Among the most popular conditions are:

Universal Domain Any combination of consistent individual judgments is in the domain of F.

Collective Rationality $F(J_1, \ldots, J_n)$ is a consistent and complete collective judgment set on F.

Anonymity For any two profiles $(J_1, \ldots, J_n)$ and $(J'_1, \ldots, J'_n)$ which are permutations of each other, $F(J_1, \ldots, J_n) = F(J'_1, \ldots, J'_n)$.

The first two conditions, Universal Domain and Collective Rationality, express that we are looking for a general aggregation procedure that outputs consistent group judgments. Anonymity articulates the egalitarian intuition: the group judgment set is invariant with respect to the "position" of an agent in the group. This is evidently inspired by democratic decision-making where votes are cast anonymously and cannot be traced back to the voter. As a consequence, the idea of experts who have special weight in the group is eliminated in favor of having a genuinely egalitarian decision model.

For judgment aggregation, this natural idea has an awkward consequence as soon as another condition, **Systematicity**, is added. That condition requires, informally, (i) that the group judgment on each proposition in the agenda depend only on individual judgments on that particular proposition, (ii) that the same aggregation criterion be used for each single proposition. In a much-cited paper, List and Pettit show that the above four criteria are jointly incompatible with each other.

Theorem 1 (Classic Impossibility Result for Judgment Aggregation) *There is no judgment aggregation function that satisfies Universal Domain, Collective Rationality, Anonymity, and Systematicity.*

(List and Pettit 2002)

Table 9.1 **The discursive dilemma for majority voting illustrates the impossibility result by List and Pettit.**

Group Member	p	$p \rightarrow q$	q
Alice	Yes	Yes	Yes
Bob	Yes	No	No
Carol	No	Yes	No
Majority Judgment	Yes	Yes	No

This impossibility result has subsequently been extended and generalized—see List and Puppe (2009) and List (2012) for introductory reviews. A classic instance of List and Pettit's impossibility result is the **discursive dilemma** for majority voting. Assume that a group of three members, Alice, Bob, and Carol, have to vote on propositions $p, p \rightarrow q$, and q. If they agree on majority voting as a judgment aggregation procedure for all three propositions, their consistent individual judgments may yield an inconsistent collective judgment. See Table 9.1 for an example.

This means that we have to part with at least one of the four above conditions—see List (2005) for an extended discussion. Giving up Universal Domain seems to make things too easy by restricting the types of disagreements that may occur in practice. Why shouldn't Alice, Bob, and Carol be allowed to have all kinds of consistent individual judgments on the agenda $\{p, p \rightarrow q, q\}$?

Abandoning Collective Rationality may look counterintuitive, but need not be a bad idea. Suspending judgments on *some* of the propositions or threshold choosing context-sensitive aggregation procedures with a supermajority threshold may avoid inconsistencies. This move may also reflect that diverse aggregation problems demand diverse solutions. Still, it would be a more principled option to dispute Systematicity or Anonymity instead.

Indeed, Systematicity may appear unreasonably strong as soon as one realizes that propositions in an agenda are often interconnected: no judgment is entirely independent of judgments on the other propositions. In addition, for some propositions, e.g., those where an error has severe practical consequences, it may be reasonable to use a different quorum than for other propositions in the agenda. Such a move would contradict Systematicity. Still, it is a delicate issue to prioritize some of the propositions over others, and good context-sensitive reasons need to be given. An epistemic justification for premise-based judgment aggregation, that is, judgment aggregation on the *reasons* for a group decision, has been provided by Bovens and Rabinowicz (2006), Hartmann, Pigozzi, and Sprenger (2010), and Hartmann and Sprenger (2012).

In favor of abandoning Anonymity, it could be said that procedural uniformity (as encoded in Systematicity) is a big practical asset. Also, some agents may possess a higher level of expertise than others. All this motivates the abandonment of Anonymity and the investigation of differential models where systematicity-like conditions can be maintained. We will examine them in the next section. However, practical constraints (e.g., democratic decision-making) often require that Anonymity be endorsed. Therefore, a solution among these lines can only be partial.

Let us now turn to the aggregation of fine-graded epistemic attitudes, such as probabilistic degrees of belief. Perhaps the impossibility results for judgment aggregation are just an artifact of the binary aggregation setting and will disappear in the more expressive probabilistic framework?

9.3 Probability Aggregation

The problem of **probability aggregation** is to reconcile probability measures over a σ-algebra $\mathcal{A}$ into a single (group) probability measure. These probabilities can be naturally interpreted as representing epistemic attitudes, that is, as individual and collective degrees of belief over the propositions in the algebra. Formally, we are looking for an aggregation function $F : \mathcal{P}^n \rightarrow \mathcal{P}$ that maps individual probability measures over $\mathcal{A}$, $(p_1, \ldots, p_n)$, to a group probability measure p^*.

A natural constraint on probability aggregation is **Convexity**, stating that the group probability of a proposition A, $p^*(A)$, should lie in between the minimum and the maximum of the individual probabilities $(p_1(A), \ldots, p_n(A))$. Therefore, some form of averaging (e.g. arithmetic or geometric, with or without weights) appears to be the natural solution to the problem of reconciling probability distributions.

Another natural constraint is the

Strong Setwise Function Property (SSFP) There is a function $g : [0,1]^n \rightarrow [0,1]$ such that for any event $A \in \mathcal{A}$:

$$F(p_1, \ldots, p_N)(A) = g(p_1(A), \ldots, p_N(A))$$

This condition requires, similar to Systematicity in judgment aggregation, that the group probability of any proposition A only depend on the individual probabilities of A, and be screened off from other propositions. SSFP directly yields the

Zero Preservation Property (ZPP) If for a proposition $A \in \mathcal{A}$ and all group members $\{1, \ldots, N\}$, $p_1(A) = \ldots = p_N(A) = 0$, then also $F(p_1, \ldots, p_N)(A) = 0$.

ZPP expresses the very natural idea that if every agent considers an event impossible (probability zero), then also the group should find it impossible.

As shown independently by McConway (1981) and Wagner (1982), any probability aggregation function that satisfies SSFP is a linear aggregation rule:

Theorem 2 (McConway 1981, Wagner 1982) *Any probability aggregation function $F : \mathcal{P}^n \rightarrow \mathcal{P}$ over an algebra $\mathcal{A}$ that satisfies SSFP is of the form*

$$F(p_1, \ldots, p_N)(A) = \sum_{j=1}^{n} \omega_j p_j(A) \tag{9.1}$$

for some weights $\omega_1, \ldots, \omega_n \in [0,1]$ such that $\sum_{j=1}^{n} \omega_j = 1$.

In other words, probability aggregation in agreement with SSFP reduces to a linear average of the individual probabilities. Note that the weights need not be equal to each other, as a purely egalitarian model would require.

This consequence of SSFP is very elegant, but it leads into trouble—like the consequences of its judgment aggregation counterpart Systematicity. Consider the following property:

Independence Preservation Whenever for two propositions $A,B \in \mathcal{A}$, we have $p_j(A \wedge B) = p_j(A) \cdot p_j(B)$ for all $1 \leq j \leq n$, then also

$$F(p_1,\ldots,p_N)(A \wedge B) = F(p_1,\ldots,p_N)(A) \cdot F(p_1,\ldots,p_N)(B)$$

In other words, if all group members agree that two propositions are probabilistically independent, then this independence should be preserved in the group judgment. Unfortunately, this reasonable property is incompatible with SSFP:

Theorem 3 (Lehrer and Wagner 1983) *There is no non-dictatorial probability aggregation function $F : \mathcal{P}^n \to \mathcal{P}$ over an algebra $\mathcal{A}$ that satisfies SSFP as well as Independence Preservation.*

In addition, linear probability aggregation (that is, aggregation in agreement with SSFP) does not commute with Bayesian Conditionalization. And a natural rule that commutes with Conditionalization, like geometric averaging, fails to satisfy other desirable properties, such as Zero Preservation and Independence Preservation—see Brössel and Eder (2013) for an overview. Similar results hold, by the way, for the aggregation of *causal judgments*, that is, the aggregation of causal (in)dependency relations represented by directed acyclical graphs (Bradley, Dietrich, and List 2014).

One of the responses to Lehrer and Wagner's impossibility results is to restrict SSFP to a subset of propositions (see Genest and Zidek (1986) for a review). This move is again parallel to the judgment aggregation literature where premise- and conclusion-based aggregation rules have been investigated. Another option is the adoption of a full Bayesian model for aggregating probability distributions (e.g., Lindley 1983).

The numerous impossibility results that exist in both domains—judgment aggregation and probability aggregation—suggest that the aggregation of epistemic attitudes on an interconnected set of propositions is just a very hard problem. It is also clear that the egalitarian presumption of judgment aggregation (Anonymity) is not the culprit for the impossibility results since similar results hold for linear aggregation procedures with differential weights. Rather, the results suggest that it is the logical connection between the propositions in the agenda that creates problems for "natural" aggregation procedures.

In the remainder, we therefore focus on aggregation procedures for a single proposition, or for estimators of a numerical quantity. This makes the impossibility results disappear, but it also leads to a shift in focus: we introduce models of differential opinion aggregation and compare their epistemic performance to egalitarian models.

9.4 Consensual Opinion Aggregation

When real groups make judgments or decisions, their members *interact* with each other: they exchange relevant information, put forward arguments and deliberate the reasons for a particular position. The previous sections did not take such interactions into account. And while it would go beyond the scope of this article to review the psychological and philosophical literature on group interactions, a particular phenomenon is relevant for our purposes: the tendency toward uniformity.

A long research tradition in social psychologists explores how groups combine individual judgments, and which decision rule (e.g., simple majority, weighted majority, "truth wins," etc.) describes the group behavior best (e.g., Lorge and Solomon 1959). In a classic study on group judgments in intellective problems (that is, problems with a high level of demonstrability), Thomas and Fink (1961) compare three different models: an **independent model**, where the group reliability is just the probability that each group member has solved the problem, a **rational model**, where the group makes a correct judgment as soon as a single member is right, and a **consensus model**, which assumes the group's inclination toward uniformity. Using an arithmetically simple, but conceptually tricky mathematical problem, the authors find that the consensus model describes the outcomes better than the other two.

Presumably inspired by these findings, Davis (1973, 122) developed the influential Social Decision Scheme (SDS) model where probability distributions that describe individual preferences are transformed into a group probability distribution over the alternatives. That is, group members make an individual judgment in terms of a probability distribution which is subsequently transformed into a group judgment, by means of a matrix multiplication procedure. Extensions of this approach to the problem of combining numerical estimates have been provided by Davis's (1996) Social Judgment Scheme, and Hinsz (1999)'s SDS-Q ("Q" standing for "quantitative") models. In these cases, weights may also be determined as a function of the *centrality* of an estimate and its distance to other estimates.

While these models are silent on the mental attitudes of the group members, they inspire the philosophical question of whether it is rational to revise one's beliefs or estimates in the light of disagreement with other group members. This question has received much attention in the recent epistemology literature on peer disagreement (e.g., Kelly 2005; Elga 2007), but it has also been applied to group judgments. In particular, philosophers have asked themselves whether it is possible to give rational foundations to consensual opinion aggregation.

The most popular idea is that **mutual respect** among the group members should prompt every group member to revise her initial opinion. This respect can be epistemically motivated (e.g., by realizing that the other group members are no less competent that oneself), but also reflect degrees of care or relations of social power, dependent on whether there is a matter of fact to the subject of disagreement. Conditional on such mutual respect, blending one's opinions with the opinions of the other group members seems to be a requirement of rationality:

> One justification for aggregation is consistency, since refusing to aggregate is equivalent to assigning everyone else a weight of zero [...]. (Lehrer and Wagner 1981, 43)

In other words, refusing to blend one's opinions would amount to unjustified *dogmatism* (see also Lehrer 1976). This argument is, by the way, independent of the question of whether the aggregation procedure should be differential or egalitarian. It just motivates the view that group rationality need not be a question of choosing the right aggregation procedure at group level, but also a question of individual belief revision.

Among respect-based models of opinion aggregation, the Lehrer-Wagner model is most prominent. It was first developed as a descriptive mathematical model of group power relations in French (1956) and as a general model of consensus formation in DeGroot (1974). While DeGroot intended his model to be normative, his research focused on the fundamental mathematical properties of the model, leaving the interpretation, further elaboration, and philosophical justification to Lehrer and Wagner (1981).

The model tackles the problem of estimating a particular quantity x, from the individual estimates v_i of every group member i. This quantity x is normally thought of as objective and independent of the group members' cognitive states. The quantity x in dispute might, for instance, be the size of the population of an endangered species, or the number of rainy days in the Netherlands in 2014.

Lehrer and Wagner's central idea consists in ascribing the agents beliefs about each other's expertise, or in other words, mutual assignments of respect as epistemic agents on the issue at hand. Then, the w_{ij} describe the proportion to which j's opinion on the subject matter in question affects i's revised opinion. These mutual respect assignments are used to revise the original estimates of the quantity in question, and codified in an $N \times N$ matrix W (where N denotes the number of agents in the group):

$$W = \begin{pmatrix} w_{11} & w_{12} & \dots & w_{1N} \\ w_{21} & w_{22} & \dots & w_{2N} \\ \dots & \dots & \dots & \dots \\ w_{N1} & w_{N2} & \dots & w_{NN} \end{pmatrix}.$$

The values in each row are nonnegative and normalized so as to sum to 1: $\sum_{j=1}^{N} w_{ij} = 1$. Thus, the w_{ij} represent relative weights which the agents ascribe to themselves and to others when it comes to estimating the unknown value x. Then, W is multiplied with a vector v that contains the agents' individual estimates of x, obtaining a novel updated value for v:

$$W \cdot v = \begin{pmatrix} w_{11}v_1 + w_{12}v_2 + \ldots + w_{1N}v_N \\ w_{21}v_1 + w_{22}v_2 + \ldots + w_{2N}v_N \\ \dots \\ w_{N1}v_1 + w_{N2}v_2 + \ldots + w_{NN}v_N \end{pmatrix}.$$

In general, this procedure will not directly lead to consensus, since the entries of $W \cdot v$ differ: $(Wv)_i \neq (Wv)_j$. However, Lehrer and Wagner (1981) show that under very weak constraints, the sequence $(W^k)_{k \in \mathbb{N}}$ converges to a matrix W^∞ where all rows are identical, that is, where all agents agree on their relative weights. That is, when the procedure of averaging is repeated, the agents will finally achieve a consensus and not only agree on the factual subject matter, but also on the differential weight that each group member should obtain.

Is the Lehrer-Wagner model a *rational* model of disagreement resolution and group decision-making? On the pro side, it has been shown that the opinion changes of the agents can be represented as a form of Bayesian updating: agents exchange information and at every step, their degrees of belief are revised by conditioning on the pieces of information they receive from their peers (Geanakopoulos and Polemarchakis 1982; Romeijn and Roy 2015). So Lehrer-Wagner opinion aggregation can, at least in principle, be made compatible with Bayesian rationality standards.

On the con side, it has been argued that Lehrer and Wagner fail to provide an epistemic story of why we should change our opinions in this peculiar way. Suppose an agent determines and normalizes her respect weights for the other group members. Then, it remains opaque why we should choose a linear updating model rather than geometric weighting or another form of averaging (Martini, Sprenger, and Colyvan 2013). Indeed, linear averaging is particularly sensitive to outliers in the individual estimates. If the group members determine the relative weights as a function of their mutual respect and independent of the submitted estimates, then extreme opinions will have an overly large impact on the group consensus.

Another worry is that respect-based differential weights may be caused by various forms of bias (Faust 1984; Trout 2009). When we assign weights based on mutual respect, biases easily distort the reasons for the weights assignment, so that there might be very little relation between the weights assigned and the objective, epistemic, weight that a certain opinion should receive. See the following section for a deeper analysis of this worry.

The most prominent objection, however, concerns the justification for *repeated* linear averaging. Lehrer and Wagner (1981) provide two stories that suggest why this might be rational. The first is a *temporal interpretation*: the disagreement after the first round of averaging is qualitatively similar to the initial disagreement. So the only way to avoid unjustified dogmatism is to repeat belief revision until consensus is reached. Wagner (1978) draws an analogy to sharing anonymous position papers after each round of aggregation: agents distribute their revised view and the reasons for that view among the other group members. This is actually similar to the famous Delphi method for structured forecasting developed by the RAND corporation (Helmer-Hirschberg 1967; Linstone and Turoff 1975): in that procedure, group members fill in questionnaires, comment on their responses and subsequently receive a (filtered) summary of the opinions of the other group members. This procedure is repeated until consensus is eventually reached.

A concern with this interpretation is that second-order expertise does not play any role. One need not be an expert oneself in order to make an accurate judgment—as long as one is able to identify which group members are competent and which aren't. If the Lehrer-Wagner model strives to be a model of group rationality, such considerations should not be neglected. However, the second interpretation of the iterated matrix multiplication that Lehrer and Wagner provide—as an amalgamation of different orders of expertise—quickly becomes cognitively absurd since one cannot rationally believe that considerations of fourth-, fifth- or sixth-order expertise should be as important as first- and second-order expertise.

For more elaborate criticism of the Lehrer-Wagner model, see Loewer and Laddaga (1985) and Martini, Sprenger, and Colyvan (2013). The latter argue that the Lehrer-Wagner model should be understood as a model of social influence and not as a genuinely epistemic model. In the view of Martini, Sprenger, and Colyvan, respect-based models of opinion aggregation are better suited for judgmental than for intellective tasks: that is, for making decisions on non-factual matters, or for aggregating preferences in a group of agents that respect each other, such as friends or colleagues. For modeling how individual *expertise* increases group accuracy, models where the weights are not determined by mutual respect, but by properly epistemic considerations, may be more adequate.

9.5 An Epistemic Analysis of Differential Weighting

The previous section has investigated differential models of opinion aggregation where the individual weights are correlated with a group member's social influence or perceived expertise. In this section, we conduct an epistemic analysis of differential weighting procedures: which relations between objective performance indicators and the relative weights have to hold to improve group performance with respect to the natural benchmark, straight averaging? This also addresses a lacuna in the Lehrer-Wagner model where such a link between truth-tracking and specific weighting schemes is not provided.

A fundamental objection to differential weighting, which we address first, is the problem of identifying experts. This question gains urgency in the light of studies that show that in spite of modest correlations, there is a substantial gap between actual and perceived expertise (e.g., Trotman et al. 1983; Littlepage et al. 1995). However, expertise may also be recognized implicitly. In this case, experts tend to exert greater influence on the group judgment than non-experts (Bonner, Baumann and Dalal 2002). This is good news because it indicates that expert-dependent decision schemes may successfully operate without the ability to make explicit such a ranking. Indeed, research on intellective tasks such as *Mastermind* and letter-number-matching demonstrates that groups tend to use expert-weighted social decision schemes for such tasks, and achieve a performance that roughly corresponds to the best individual member (Bonner 2004; Baumann and Bonner 2004). This sounds modest, but actually, it is a substantial achievement if we do not know in advance who the experts are (see also Laughlin and Ellis 1986; Libby, Trotman, and Zimmer 1987; Laughlin, Hatch, et al. 2006; Bonner, Silito, and Baumann 2007). In this context, it is notable that performance feedback does not substantially help the agents to recognize expertise and to improve performance. Anyway, these findings indicate that differential weighting can be epistemically beneficial in a variety of contexts, and that expert recognition is no practically impossible task. This brings us to the theoretical question of *how* we should weigh the experts.

A classical result in this area concerns the **aggregation of binary forecasts**. Assume that we have to predict whether it will rain on the next day. The agents are conceptualized as a group of independent forecasters with a certain probability p_i of getting the result right. How should these forecasts be combined? Nitzan and Paroush (1982) and Shapley and Grofman (1984) show that if rain is a priori as likely as no rain, group accuracy is maximized by following a weighted majority voting rule where the weights of the agents are proportional to their logarithmic betting odds:

$$w_i \propto \log \frac{p_i}{1-p_i} \qquad (9.2)$$

This result is, in fact, closely related to Bayes' theorem. However, it only solves the problem of **combining binary forecasts** and does not address the more general problem of **combining numerical estimates** of an unknown quantity μ. These estimates can correspond to the individual opinions of a group of agents, but also to the outputs of different mathematical models, e.g., different predictions for the extent of global warming. Our problem consists in finding a method of combining these estimates in an advantageous way, and stating general conditions for when taking into account individual expertise improves the group judgment.

We address this question through a simple statistical model developed by Klein and Sprenger (2015). Their work builds on analytical work in the forecasting and social psychology literature (Bates and Granger 1969; Hogarth 1978), following the approach of Einhorn, Hogarth, and Klempner (1977). Because of its simplicity and generality, the Klein-Sprenger model is especially well suited for principled comparisons of egalitarian and differential opinion aggregation. It also stands in a venerable research tradition in social psychology: agents are modeled as independent signalers with a certain reliability (e.g., Zajonc and Smoke 1959). Such formal models are then used as a standard for gauging empirical findings, and they may indicate how information should be spread over the agents in order to optimize performance in a recall task or similar cognitive problems.

Klein and Sprenger model the group members' individual estimates X_i, $i \leq n$, as independent random variables that scatter around the true value $\mu = 0$ with zero bias and variance σ_i^2. No further distributional assumptions are made in order to preserve the generality of the analysis. The competence of an agent (or scientific model) is explicated as the degree of precision in estimating the true value. Then, the epistemic question about the epistemic benefits of differential weighting can be translated into a precise mathematical question:

> **Problem:** Which convex combination of estimates $\hat{\mu} = \sum_{i=1}^{n} c_i X_i$ should the agents choose in order to reduce expected square loss?

This mathematical question serves to identify a modeling target which is reasonably close to the actual problem and which we can use to study the epistemic properties of differential weighting in opinion aggregation.

It is well-known that for any such estimate $\hat{\mu}$, the mean square error (MSE) can be calculated as

$$
\begin{aligned}
\mathrm{MSE}(\hat{\mu}) = \mathbb{E}[(\hat{\mu} - \mu)^2] &= \mathbb{E}\left[\left(\sum_{i=1}^{n} c_i X_i\right)^2\right] \\
&= \sum_{i=1}^{n} c_i^2 \mathbb{E}\left[X_i^2\right] + \sum_{i \neq j} c_i c_j \mathbb{E}[X_i]\mathbb{E}[X_j] \\
&= \sum_{i=1}^{n} c_i^2 \sigma_i^2 \qquad (9.3)
\end{aligned}
$$

which is minimized by the following assignment of the c_i (cf. Lehrer and Wagner 1981, 139):

$$c_i^* = \left(\sum_{j=1}^{n} \frac{\sigma_i^2}{\sigma_j^2} \right)^{-1}. \tag{9.4}$$

The problem with these optimal weights is that each agent's individual expertise would have to be known in order to calculate them. They can also be quite extreme. Given all the biases that actual deliberation is loaded with, e.g., ascription of expertise due to professional reputation, age or gender, or bandwagon effects, it is unlikely that the agents succeed at unraveling the expertise of all other group members to such a precise degree (cf. Nadeau, Clouthier, and Gray 1993; Armstrong 2001). In line with what has been said before, it is more realistic to expect that groups may be *qualitatively* competent at identifying experts, but not at determining the optimal weights.

Therefore the scope of the inquiry is widened:

> **Question:** Under which conditions will differentially weighted group judgments outperform the straight average?

A first answer is given by the following results, where the differential weights preserve the expertise ranking (see Klein and Sprenger (2015) for all results and proofs):

Theorem 4 (First Baseline Result) *Let $0 \leq c_1 \leq \ldots \leq c_n \leq 1$ be the weights of the individual group members, that is, $\sum_{i=1}^{n} c_i = 1$. Further assume that for all $i > j$:*

$$1 \leq \frac{c_i}{c_j} \leq \frac{c_i^*}{c_j^*} \tag{9.5}$$

Then the differentially weighted estimator $\hat{\mu} := \sum_{i=1}^{n} c_i X_i$ outperforms the straight average. That is, $MSE(\hat{\mu}) \leq MSE(\bar{\mu})$, with equality if and only if $c_i = 1/n$ for all $1 \leq i \leq n$.

This result demonstrates that knowledge of the exact competence of agents is not required for improving decisions with respect to the straight average baseline. Rather, as long as the competence is ranked in the right order, the differentially weighted estimate will outperform the straight average.

The following result extends this finding to a case where the benefits of differential weighting are harder to anticipate: we allow the c_i to lie in the entire $[1/n, c_i^*]$ interval, allowing for cases where the ranking of the group members is not represented correctly. One might conjecture that this phenomenon adversely affects performance, but this is not the case:

Theorem 5 (Second Baseline Result) *Let $c_1 \ldots c_n \in [0,1]$ be such that $\sum_{i=1}^{n} c_i = 1$. In addition, let $c_i \in [\frac{1}{n}; c_i^*]$ (respectively $c_i \in [c_i^*; \frac{1}{n}]$) hold for all $1 \leq i \leq n$. Then the differentially weighted estimator $\hat{\mu} := \sum_{i=1}^{n} c_i X_i$ outperforms the straight average. That is,* $\text{MSE}(\hat{\mu}) \leq \text{MSE}(\bar{\mu})$*, with equality if and only if $c_i = 1/n$ for all $1 \leq i \leq n$.*

In other words, as long as the relative weights lie in between the equal weights and the optimal weights, the accuracy of the group judgment is increased. Even a fallacious competence ranking need not be harmful: the resulting estimate will still be better than straight averaging. Briefly, as long as there is a positive correlation between degrees of expertise and impact on the group judgment, the group does well to weigh the estimates differentially.

The litmus test for Klein and Sprenger's results are cases where some of their idealizing assumptions fail, e.g., independence or unbiasedness. For example, training, experience, risk attitude, or personality structure may bias the agents' estimates into a certain direction. In assessing the impact of industrial development on a natural habitat, an environmentalist will usually come up with an estimate that significantly differs from the estimate submitted by an employee of a corporation that intends to exploit the habitat—even if both are intellectually honest and share the same information. In these circumstances, the agents should not be modeled as unbiased statistical estimators, but as estimators whose mean value is different from μ. However, as long as the differentially weighted bias is smaller or equal than the average bias, the baseline results remain valid and differential weighting still outperforms straight averaging (Section 3 in Klein and Sprenger 2015).

Consider now the case where agents are not independent, but where their opinions are correlated with each other, e.g., because they draw from similar information sources (e.g., Goldman 2001). This may happen because they use similar research methods or because they share information with each other. For this case, Klein and Sprenger show the following result:

Theorem 6 *Let $X_1,\ldots,X_n$ be unbiased estimators, that is, $\mathbb{E}[X_i] = \mu = 0$, and let the c_i satisfy the conditions of one of the baseline results, with $\hat{\mu}$ defined as before. Let $I \subseteq \{1,\ldots,n\}$ be a subset of the group members with the property*

$$\forall i,j \neq k \in I : c_i \geq c_j \;\Rightarrow\; \mathbb{E}[X_j X_k] \geq \mathbb{E}[X_i X_k] \geq 0. \tag{9.6}$$

(i) ***Correlation vs. Expertise*** *If $I = \{1,\ldots,n\}$, then weighted averaging outperforms straight averaging, that is,* $\mathrm{MSE}(\hat{\mu}) \leq \mathrm{MSE}(\bar{\mu})$.

(ii) ***Correlated Subgroup*** *Assume that $\mathbb{E}\left[X_i X_j\right] = 0$ if $i \in I$ and $j \notin I$, and that*

$$\frac{1}{|I|}\sum_{i\in I} c_i \leq \frac{1}{n}\sum_{i=1}^{n} c_i. \tag{9.7}$$

Then weighted averaging still outperforms straight averaging, that is, $\mathrm{MSE}(\hat{\mu}) \leq \mathrm{MSE}(\bar{\mu})$.

To fully understand this theorem, we have to clarify the meaning of condition (9.6). Basically, it says that in group I, experts are less correlated with other (sub)group members than non-experts.

Once we have understood this condition, the rest is straightforward. Part (i) states that if I equals the entire group, then differential weighting has an edge over averaging. That is, the benefits of expertise recognition are not offset by the perturbations that mutual dependencies may introduce. Arguably, the generality of the result is surprising since condition (9.6) is quite weak. Part (ii) states that differential weighting is also superior whenever there is no correlation with the rest of the group, and as long as the average competence in the subgroup is lower than the overall average competence (see equation (9.7)).

It is a popular opinion (e.g., Surowiecki 2004) that correlation of individual judgments is one of the greatest dangers for relying on experts in a group. To some extent, this opinion is reflected by the above theorem. However, expertise-informed group judgments may still be superior to straight averaging, as demonstrated by Theorem 6. Thus, the interplay of correlation and expertise is subtle and cannot be generalized easily.

Summing up, taking into account relative accuracy positively affects the epistemic performance of groups even if the ranking of experts is only partially reliable (Theorems 4 and 5). The result remains stable over several representative extensions of the model, such as various forms of bias, violations of independence, and over- and underconfident agents (e.g., Theorem 6). In particular, differential weighting is superior (i) if experts are, on average, less biased; (ii) if all agents share the same sort of bias; (iii) if experts are less correlated with the rest of the group than other group members. These properties may be surprising and demonstrate the stability and robustness of expertise-informed judgments, implying that the benefits of recognizing experts may offset the practical problems linked with that process. The parsimony of this model and the independence of specific distributional assumptions suggest that these qualitative phenomena are likely to occur in reality, too.

9.6 Expert Judgment Literature

While the preceding sections dealt with various formal models for describing opinion aggregation and group judgments, this section surveys a limited number of empirical results regarding the elicitation and practical use of individual expertise. In particular, we present empirical justifications for both egalitarian and differential strategies in opinion aggregation.

9.6.1 Justification of Egalitarian Approaches

The empirical literature on expert judgment reflects the divisions and problems of the formal literature. On the one hand, scholars have supported equal weighting and independent forecasts on grounds that giving more weight to a method or another is unjustified unless the results are already known. Let us take an example: Imagine that an economist and a political scientist are forecasting the next 50 years of growth-rates for China. The two scientists may not only use different prediction strategies, but also focus on different sources of evidence: The economist may be inclined to use model-projections from past time-series, while the political scientist may rely on his personal rules-of-thumb, and "eyeball" her estimates based on intuitions about likely historical trends of nations with an economic history similar to today's China. Who is to be weighted more?

Which of the two strategies is the most successful one cannot be known a priori. If the world stays more or less the same, we may expect model-projections to be more accurate than eyeballing. But time-series projections suffer from known "broken-leg problems" (Bishop and Trout 2005, 45–53); that is, they ignore data that is particularly disruptive of otherwise smooth time-series. A political economist might be inclined to take China's aging problem more seriously than someone who looked only at index numbers; for instance, as a warning of the fact that China may well hit a growth-rate plateau in the next few decades.

Typically, in the selection of expertise we can nonetheless rely on the fact that experts are already selected with respect to their past performance. So the sociologist might be preferred over the economist, or vice versa, on grounds that she has forecast past geopolitical events more successfully. We could, that is, give differential weights on the basis of past performance. This strategy too has its own flaws. In order for past performance to be meaningful—my judgment is as likely to be correct in the future as it has been in the past—the class of forecasts we are considering must be the relevant one. But it is often difficult to know which class of forecasts is the relevant one for the kind of problem we are considering. Kitcher offers a good example of the problem of relevance: In a thorough reconstruction of the disagreements over climate change, Oreskes and Conway (2010) show how several climate change skeptics had in fact obtained their status of experts in fields other than those relevant to climate sciences (see Kitcher 2010).

Consider the following simplified illustration. Let us assume that the political economist, in the example above, is a very reliable long-term forecaster of geopolitical events—she has successfully forecasted the geopolitical situation of several countries many years in advance. The economist, on the other hand, can very reliably predict the trend of key economic indicators (GDP, inflation, public debt, etc.). On the one hand, when purely economic factors are more likely to affect China's GDP growth in the next 50 years—that is, assuming a relative stability of the geopolitical system—then the economist has probably a better track record than the political scientist. On the other hand, if we think that geopolitical events are more likely to shape the future of Chinese growth, then the political scientist will likely have a better shot at the correct forecast. What the example highlights is that the two track records of individual experts are not always comparable—see Reiss (2008, 38–41) and Martini (2014) for a more detailed discussion.

In sum, selecting experts on grounds of performance indicators, credentials, etc. can be like comparing apples and oranges. There are often no one-dimensional meters of comparison with which to assess expertise. Worse still, a scale can be handpicked by a relevant interest group in order to favor a preferred (biased) outcome as the *right* outcome. For the foregoing reasons, whenever we do not have a clear picture of the problem, of the kind of expertise that is relevant to it, etc., then we have an argument for egalitarian aggregation mechanisms, that is: equal-weighted averaging. Equal-weighted averaging is discussed in extensively in Armstrong (2001) and typically used in the Delphi method for aggregating opinions (see Dalkey 2002). In his *Principles of Forecasting* Armstrong recommends using "equal weights unless you have strong evidence to support unequal weighting of forecasts" (2001, 422). He refers to Clemen, who "conducted a comprehensive review of the evidence and found equal weighting to be accurate for many types of forecasting" (see Clemen 1989). However, Armstrong notices, "the studies that [Clemen] examined did not use domain knowledge" (2001, 422). It is on the caveat of 'domain knowledge' that differential aggregation models can, under certain conditions, be preferable to egalitarian ones, as the next section will illustrate.

Before we discuss the case of differential weighting, it must be noted that the phase of expert judgment aggregation logically comes after one has identified the experts in the first place. While aggregating judgment can, to a great extent, be reduced to a formal process, the process of separating experts from laypeople is instead a highly empirical matter. Experts are usually identified by a number of cognitive and non-cognitive traits that typically correlate with expertise. For example, experts tend to be more consistent than laypeople in their judgment (see Shanteau 2002), they have more problem-discrimination ability, being able to see more subtleties in otherwise similar problems than laypeople (see Hammond 1996), and they can usually display a more extensive track record of experiences in their field of expertise (see Goldman 2001). The traits just mentioned are just a few of those usually associated with expertise. While we can only touch briefly on this matter here, the reader can find extensive discussions on how to tell experts apart from laypeople in Walton (1989), Goldman (2001), Shanteau (1992, 2002), Reiss (2008), and Martini (2014, 2015).

9.6.2 Justification of Differential Approaches

In some cases using differential weighting is justified by the fact that some of the members in the group possess more knowledge than others, and are therefore more likely to give accurate judgments. This typically happens in contexts where *domain knowledge* is involved. Domain knowledge is knowledge that is typical of a specific subject, field of research, etc., and is therefore most likely in possession of those with advanced training or experience in that field. To give an example, forecasting the trend of an economic indicator by means of an econometric model involves domain-specific knowledge of statistics and econometric modeling; unlike the case of forecasting general geopolitical trends, which involves domain-general, rather than domain-specific, knowledge. But let us take a different case. Cooke (1991, 159ff.) reports on an experiment conducted at a Dutch training facility for operators of large technological systems. The experiment involved highly trained professionals, and aimed at testing whether experience is correlated with calibration. Subjects were tested over general knowledge questions and domain-specific questions. Following are some examples of the two types.[2]

- General knowledge questions
  - What was the total egg production in millions in the U.S. in 1965?
  - What is the total number of students currently enrolled in the Doctoral Program at the Harvard Business School?
- Domain knowledge questions
  - What is the maximal efficiency of the Tyne RM1A gas turbine?
  - What is the maximum admissible intake temperature for gas in the Olympus power turbine?

[2] The examples are taken from Cooke (1991) and from Alpert and Raiffa (1982), the latter being the source of some of Cooke's own experimental questionnaires.

The distinction between general knowledge and domain knowledge is not a clear-cut one, leaving much room for debate in places where the two might overlap considerably. Nonetheless, we can safely say that general knowledge questions can usually be answered by reasoning on items of information that are widely shared. When trying to answer the question of how many students are currently enrolled at Harvard's MBA, we might be helped by reasoning over the following questions: How many students are typically enrolled in a university? How many in a typical American university? How large the Harvard Business School can be with respect to a typical business school? What is the minimum size of a business school to operate efficiently? So reasoning over related questions is likely to take us close to a reasonable range—i.e., the Harvard Business School has between 1000 and 5000 students.

Unlike general knowledge, domain knowledge involves, in the first place, a terminology that is only mastered with specific education or training. The average educated person probably has very little information on how to even express the efficiency of a turbine, viz. "what are the units?" It is even less likely that we could find anyone who is not versed in the field of engineering, or more specifically aerospace engineering, who possesses the relevant information for estimating the admissible intake temperature of a turbine, or the efficacy of a specific model of turbine like the Tyne RM1A (cf. above).

Domain knowledge problems are such that we can often make a distinction between "experts" and "laymen," and for which we can hope to find sensible differential weights to be used in aggregation. The two important question then remain, from an empirical perspective, *when* to use differential aggregation, as opposed to equal weights, and, if used, how to find appropriate weights. The formal literature can help us find aggregation rules that respect a number of formal desiderata. But empirical assessment is just as important in discovering the efficiency of different weighting schemas.

It should be noted here that there are two general strategies for selecting weights: *ex-ante* weighting and *ex-post* weighting (see Ashton and Ashton 1985). The former strategy looks at reasons for choosing a certain weighting scheme based on evidence that is anterior to the results of the weighting. For instance, a manipulable weighting function that excludes (or reduces the importance of) some experts based on irrelevant evidence may be discarded a priori. But choosing an ex-ante weighting scheme may not be enough: we may want to calibrate the weights by an empirical (i.e., ex-post) assessment of the weighting scheme. An influential a posteriori strategy for selecting weights is given in Cooke (1991): weights are found by averaging an expert's performance on a number of seed questions in his or her own field of expertise, and relative to the field of expertise that is deemed relevant to the problem at hand. Seed questions are selected among knowledge items that are relevant for the problem in which the facilitator is interested, but whose answers are not known in advance by the agent that is being assessed. A seed problem could be, for example, "estimate the probability of failure of the Tyne RM1A gas turbine under a certain level of mechanical stress."

In Cooke's methodology, weights relative to expert *X* are determined on the basis of expert *X*'s performance on a "quiz" that the facilitator thinks is relevant for the problem she is interested in. An engineer who is asked to assess the risks related to a nuclear power plant will be assessed (and weighted) on the basis of seed questions related, for instance, to components of nuclear power plants, the materials used in the contraction of power plants, etc. There is clearly an extrapolation problem here: The seed questions, it is assumed, are indicative of expertise in the problem of interest. This is clearly something that needs to be evaluated case-by-case, when formulating seed questions, and there seems to be no a priori method for deciding which seed questions are valuable and which ones are not. While it is illustrative of how we can assess a weighting schema ex-post, Cooke's method is limited in that it can only be applied in cases where the problem is defined at a very detailed level, and where relevant seed questions are available. The burden of proof is therefore on those who want to use differential weighting to prove their case (see also Armstrong 2001).

9.7 Conclusions

This article has given a survey over the problem of combining individual judgments into group judgments, with a distinct focus on individual expertise and differential weighting.

Our survey revealed that there are different conceptions of group rationality in opinion aggregation, and that their appropriate use depends on the context. For example, in democratic decision-making, egalitarian weighting schemes are usually compulsory, whereas in scientific contexts, considerations of expertise or relevant experience (e.g., in medical diagnosis) may be more forceful and motivate the use of a differential weighting scheme. The epistemic benefits and drawbacks of such schemes have been investigated theoretically in Section 9.5 and empirically in Section 9.6.

It also transpired that there is not necessarily an all-encompassing account of group rationality. The impossibility results in Sections 9.2 and 9.3 show that for a logically interconnected agenda of propositions, there may be no opinion aggregation procedure that satisfies a set of plausible and intuitive constraints. These results hold for the aggregation of binary judgments as well as for the aggregation of graded attitudes, e.g., degrees of belief. This was actually one of the reasons why we have investigated the problem of aggregating a single numerical estimate in Sections 9.4 and 9.5. While Sections 9.4 investigated the rationality of belief revision and reaching a consensus, Section 9.5 gave a general epistemic analysis of differential weighting schemes, albeit in a simple statistical model.

It is a typical feature of the literature on opinion aggregation and expertise that there is a variety of approaches spread over different disciplines and research methods: the same problem may be tackled from the point of view of social choice theory, (formal) epistemology, mathematics and statistics, economics, experimental psychology, management science, and risk studies. Therefore our survey necessarily remains incomplete. However, we hope to have given the reader a taste of the diversity of the approaches to modeling individual expertise, as well as of the enormously complex interplay between formal models of group judgments and empirical studies. In particular, we hope to have created a better understanding of the differences between egalitarian and differential models of opinion aggregation and group decision-making, and of the philosophical rationales behind them.

Acknowledgments

The authors thank the editors of this book and two anonymous referees for their useful feedback. Jan Sprenger wishes to thank the Netherlands Organisation for Scientific Research (NWO) for support of his research through Vidi grant #276-20-023, and the European Research Council (ERC, Starting Investigator Grant #640638).

References

Alpert, Marc, and Howard Raiffa (1982): "A progress report on the training of probability assessors," in: Amos Tversky, Paul Slovic, and Daniel Kahneman (eds.), *Judgment Under Uncertainty: Heuristics and Biases*. Cambridge: Cambridge University Press.

Armstrong, J. Scott (2001): "Combining forecasts", in: J. Scott Armstrong (ed.), *Principles of Forecasting: A Handbook for Researchers and Practitioners*. Norwell, MA: Kluwer Academic Publishers.

Ashton, Alison Hubbard, and Robert H. Ashton (1985): "Aggregating subjective forecasts: Some empirical results", *Management Science* 31(12): 1499–1508.

Bates, J.M., and C.W.J. Granger (1969): "The combination of forecasts", *Operational Research Quarterly* 20: 451–468.

Baumann, Michael R., and Bryan L. Bonner (2004): "The effects of variability and expectations on utilization of member expertise and group performance", *Organizational Behavior and Human Decision Processes* 93: 89–101.

Beatty, J. (2006): "Masking disagreement among experts", *Episteme* 3(1–2): 52–67.

Beatty, J., and A. Moore (2010): "Should we aim for consensus", *Episteme* 7(3): 198–214.

Bishop, Michael A., and John D. Trout (2005): *Epistemology and the Psychology of Human Judgment*. Oxford: Oxford University Press.

Bonner, Bryan L. (2004): "Expertise in group problem solving: Recognition, social combination, and performance", *Group Dynamics: Theory, Research, and Practice* 8: 277–290.

Bonner, Bryan L., Michael R. Baumann, and Reeshad S. Dalal (2002): "The effects of member expertise on group decision-making and performance", *Organizational Behavior and Human Decision Processes* 88: 719–736.

Bonner, Bryan L., Sheli D. Silito, and Michael R. Baumann (2007): "Collective estimation: Accuracy, expertise and extroversion as sources of intra-group influence", *Organizational Behavior and Human Decision Processes* 103: 121–133.

Bovens, Luc, and Wlodek Rabinowicz (2006): "Democratic answers to complex questions: An epistemic perspective", *Synthese* 150: 131–153.

Bradley, Richard, Franz Dietrich, and Christian List (2014): "Aggregating causal judgments", *Philosophy of Science* 81: 491–515.

Brössel, Peter, and Anna-Maria A. Eder (2014): "How to resolve doxastic disagreement", *Synthese* 191: 2359–2381.

Christensen, David (2009): "Disagreement as evidence: The epistemology of controversy", *Philosophy Compass* 4: 756–767.

Clemen, Robert T. (1989): "Combining forecasts: A review and annotated bibliography", *International Journal of Forecasting* 5: 559–583.

Cooke, Roger M. (1991): *Experts in Uncertainty*. Oxford: Oxford University Press.

Dalkey, Norman C.(2002): "A Delphi study of factors affecting the quality of life", in: Harold A. Linstone and Murray Turoff (eds.), *The Delphi Method: Techniques and Applications*. Boston: Addison-Wesley.

Davis, James H. (1973): "Group decision and social interaction: A theory of social decision schemes", *Psychological Review* 80: 97–125.

Davis's, James H. (1996): "Group decision making and quantitative judgments: A consensus model", in: E. Witte and J.H. Davis (eds.), *Understanding Group Behavior: Consensual Action by Small Group*, 35–59. Mahwah, NJ: Erlbaum.

DeGroot, Morris (1974): "Reaching a consensus", *Journal of the American Statistical Association* 69: 118–121.
Dietrich, Franz, and Christian List (2008): "Judgment aggregation without full rationality", *Social Choice and Welfare* 31: 15–39.
Einhorn, Hillel J., Robin M. Hogarth, and Eric Klempner (1977): "Quality of group judgment", *Psychological Bulletin* 84: 158–172.
Elga, Adam (2007): "Reflection and disagreement", *Noûs* 41: 478–502.
Faust, David (1984): *The Limits of Scientific Reasoning.* Minneapolis: University of Minnesota Press.
French, John R.P. Jr. (1956): "A formal theory of social power", *Psychological Review* 63: 181–194.
Geanakopoulos, John D., and Heraklis M. Polemarchakis (1982): "We cannot disagree forever", *Journal of Economic Theory* 28: 192–200.
Genest, Christian, and James V. Zidek (1986): "Combining probability distributions: A critique and an annotated bibliography", *Statistical Science* 1: 114–135.
Goldman, Alvin I. (1999): *Knowledge in a Social World.* Oxford: Oxford University Press.
Goldman, Alvin I. (2001): "Experts: Which ones should you trust?", *Philosophy and Phenomenological Research* 63: 85–110.
Grossi, Davide (2009): "Unifying preference and judgment aggregation", *AAMAS* 1: 217–224.
Hammond, Kenneth R. (1996): *Human Judgment and Social Policy: Irreducible Uncertainty, Inevitable Error, Unavoidable Injustice*. Oxford: Oxford University Press.
Hartmann, Stephan, Gabriella Pigozzi, and Jan Sprenger (2010): "Reliable methods of judgment aggregation", *Journal for Logic and Computation* 20: 603–617.
Hartmann, Stephan, and Jan Sprenger (2012): "Judgment aggregation and the problem of tracking the truth", *Synthese* 187: 209–221.
Helmer-Hirschberg, Olaf (1967): "Analysis of the future: The Delphi method". Working paper of the RAND Corporation.
Hinsz, Verlin B. (1999): "Group decision making with responses of a quantitative nature: The theory of social decision schemes for quantities", *Organizational Behavior and Human Decision Processes* 80: 28–49.
Hogarth, Robin M. (1978): "A note on aggregating opinions", *Organizational Behavior and Human Performance* 21: 40–46.
Kelly, Thomas (2005): "The epistemic significance of disagreement", in: J. Hawthorne and T. Szabo (eds.), *Oxford Studies in Epistemology*, 167–196. Oxford: Oxford University Press.
Kitcher, Philip (2010): "The climate change debates", *Science* 328: 1230–1234.
Klein, Dominik, and Jan Sprenger (2015): "Modeling individual expertise in group judgments", Klein, Dominik, and Jan Sprenger. "Modelling individual expertise in group judgements." *Economics and Philosophy* 31.01: 3–25.
Kuhn, Thomas S. (1962): *The Structure of Scientific Revolutions*. Chicago: University of Chicago Press.
Kuhn, Thomas S. (1977): *The Essential Tension*. Chicago: University of Chicago Press.
Laughlin, Patrick R., and Alan L. Ellis (1986): "Demonstrability and social combination processes on mathematical intellective tasks", *Journal of Experimental Social Psychology* 22: 177–189.
Laughlin, Patrick R., Erin C. Hatch, Jonathan S. Silver, and Lee Boh (2006): "Groups perform better than the best individuals on letters-to-numbers problems: Effects of group size", *Journal of Personality and Social Psychology* 90: 644–651.
Lehrer, Keith (1976): "When rational disagreement is impossible", *Noûs* 10: 327–332.
Lehrer, Keith, and Carl Wagner (1981): *Rational Consensus in Science and Society*. Dordrecht: Reidel.
Lehrer, Keith, and Carl Wagner (1983): "Probability amalgamation and the independence issue: A reply to Laddaga", *Synthese* 55: 339–346.
Libby, Robert, Ken T. Trotman, and Ian Zimmer (1987): "Member variation, recognition of expertise and group performance", *Journal of Applied Psychology* 72: 81–87.
Lindley, Dennis V. (1983): "Reconciliation of probability distributions", *Operations Research* 31: 866–880.
Linstone, Harold A., and Murray Turoff (1975): *The Delphi Method: Techniques and Applications*. Reading, MA: Addison-Wesley.

List, Christian (2005): "Group knowledge and group rationality: A judgment aggregation perspective", *Episteme* 2: 25–38.
List, Christian (2012): "The theory of judgment aggregation: An introductory review", *Synthese* 187: 179–207.
List, Christian, and Philip Pettit (2002): "Aggregating sets of judgments: An impossibility result", *Economics and Philosophy* 18: 89–110.
List, Christian, and Clemens Puppe (2009): "Judgment aggregation: A survey", in: Paul Anand, Prasanta Pattanaik, and Clemens Puppe (eds.), *The Handbook of Rational and Social Choice*, 457–482. Oxford: Oxford University Press.
Littlepage, Glenn E., Greg W. Schmidt, Eric W. Whisler, and Alan G. Frost (1995): "An input-process-output analysis of influence and performance in problem-solving groups", *Journal of Personality and Social Psychology* 69: 877–889.
Loewer, Barry, and Robert Laddaga (1985): "Destroying the consensus", *Synthese* 62: 79–95.
Lorge, Irving, and Herbert Solomon (1959): "Individual performance and group performance in problem solving related to group size and previous exposure to the problem", *The Journal of Psychology* 48: 107–114.
Martini, Carlo (2014): "Experts in science: A view from the trenches", *Synthese* 191: 3–15.
Martini, Carlo (2015): "The paradox of proof and scientific expertise", *Humana.Mente. Journal of Philosophical Studies* 28(3–4): 1–16.
Martini, Carlo, Jan Sprenger, and Mark Colyvan (2013): "Resolving disagreement through mutual respect", *Erkenntnis* 78: 881–898.
McConway, K. J. (1981): "Marginalization and linear opinion pools", *Journal of the American Statistical Association* 76: 410–414.
Nadeau, Richard, Edouard Clouthier, and J.-H. Gray (1993): "New evidence about the existence of a bandwagon effect in the opinion formation process", *International Political Science Review* 14: 203–213.
Nitzan, Shmuel, and Jacob Paroush (1982): "Optimal decision rules in uncertain dichotomous choice situations", *International Economic Review* 23(2): 289–297.
Oreskes, Naomi, and Erik M. Conway (2010): *Merchants of Doubt*. London: Bloomsbury Publishing.
Page, Scott E. (2007): *The Difference*. Princeton: Princeton University Press.
Reiss, Julian (2008): *Error in Economics: Towards a More Evidence-Based Methodology*. New York: Routledge.
Romeijn, Jan Willem, and Olivier Roy (2015): "All agreed: Aumann meets DeGroot". Unpublished manuscript.
Shanteau, J. (1992): "The psychology of experts: An alternative view", in: G. Wright and F. Bolger (eds.), *Expertise and Decision Support*, New York: Plenum Press. 11–23.
Shanteau, J., et al. (2002): "Performance-based assessment of expertise: How to decide If someone Is an expert or not", *European Journal of Operational Research* 136(2): 253–263.
Shapley, Lloyd, and Bernard Grofman (1984): "Optimizing group judgmental accuracy in the presence of interdependencies", *Public Choice* 43: 329–343.
Surowiecki, James (2004): *The Wisdom of the Crowds*. Harpswell: Anchor.
Thomas, Edwin J., and Clifton F. Fink (1961): "Models of group problem solving", *The Journal of Abnormal and Social Psychology* 63: 53–63.
Trotman, K.T., P.W. Yetton, and I.R. Zimmer (1983): "Individual and group judgments of internal control systems", *The Journal of Accounting Research* 21: 286–292.
Trout, J.D. (2009): *The Empathy Gap: Building Bridges to the Good Life and the Good Society*. New York: Viking/Penguin.
Wagner, Carl (1978): "Consensus through respect: A model of rational group decision-making", *Philosophical Studies* 34: 335–349.
Wagner, Carl (1982): "Allocation, Lehrer models, and the consensus of probabilities", *Theory and Decision* 14: 207–220.
Walton, N. D. (1989): "Reasoned use of expertise in argumentation", *Argumentation* 3: 59–73.
Zajonc, Robert B., and William H. Smoke (1959): "Redundancy in task assignments and group performance", *Psychometrika* 24: 361-369.

NAMES INDEX

SUBJECT INDEX

Figures and tables are indicated by "f" and "t" following the page number

www.ingramcontent.com/pod-product-compliance
Ingram Content Group UK Ltd.
Pitfield, Milton Keynes, MK11 3LW, UK
UKHW021134160926
475595UK00020B/41

9 780190 680534